Technology
in the Garden

Technology in the Garden

Research Parks and

Regional Economic

Development

Michael I. Luger & Harvey A. Goldstein

The University of North Carolina Press

Chapel Hill & London

© 1991
The University of
North Carolina Press
All rights reserved

Library of Congress
Cataloging-in-Publication Data
Luger, Michael I. (Michael Ian)
   Technology in the garden : research parks
and regional economic development / by
Michael I. Luger and Harvey A. Goldstein.
      p.   cm.
   Includes bibliographical references and
   index.
   ISBN 0–8078–2000–8 (alk. paper).—
   ISBN 0–8078–4345–8 (pbk. : alk. paper)
   1. Research parks—United States—Case
studies.   I. Goldstein, Harvey, 1947–   .
II. Title.
HC110.R4L84   1991
607'.2—dc20   91–50255
                     CIP

The paper in this book meets the guidelines
for permanence and durability of the
Committee on Production Guidelines for
Book Longevity of the Council on Library
Resources.
Manufactured in the United States of America
95 94 93 92 91
5 4 3 2 1

For

Laura and Meredith

Contents

Tables and Figures

Tables

Figures

Preface

Research parks have become a prominent element in state and regional development strategies in the United States, as well as in Western Europe and Japan, Australia, and many other developed countries. Also referred to as *science parks* and *technology parks*, they generally are intended to serve as a seedbed or catalyst for the development of a concentration of innovation- and technology-oriented business enterprises in a region or a state. In that sense, research parks are closely related in function to *science cities*, or technopoles, which are also becoming popular in some countries.

The widely accepted premise underlying the research park strategy is that a region's long-term economic viability will depend on its ability to generate and sustain a concentration of businesses capable of developing new products (or processes) that can penetrate international markets. For regions faced with a high concentration of older, declining manufacturing sectors, research parks have been viewed as a tool for facilitating economic restructuring. For other regions whose economies have been performing well, investments in research parks may represent a long-term insurance policy. In either case, the R&D-led economic development strategy, when successful, almost always leads to more than just employment growth and new business formation. It brings with it concomitant changes in occupational mix, wage, and salary structure; political culture; and spatial patterns of development. While many of these changes represent net benefits to the community and region, some of these can cause stress, particularly for older residents and members of the labor force who are employed in the traditional sectors. More generally, the benefits and costs of the induced economic development tend not to be shared equally among population groups.

Unlike Japan, France, and the Netherlands, for example, where central governments have played major roles in the creation and coordination of research parks and technopoles, the federal government in the United States has been involved only peripherally in research park development. The role of the federal government in subnational economic development policy-making in general has waned since the early 1980s. State and local governments have had to fill the void in policy-making responsibility. But they also have had to bear a much larger portion of the fiscal responsibility for economic development initiatives. As a result, many states have had to adopt an entrepreneurial and strategic approach to economic development, unlike the federal grants-based approach of the 1960s and 1970s. Research parks represent both symbolic and substantive means of attempting to increase a region's "creativity" and innovative capacity. These factors together help explain the wide popularity of research parks and the fact that the large majority of research parks now in existence have been created since 1982.

As is often the case with economic development initiatives, however, especially those originating at the state and local levels, the implementation of research parks has preceded careful evaluation. The motivations for this study come from both academic research and public policy perspectives.

As researchers in the field of regional development in a university that is closely tied to the Research Triangle Park (RTP), one of the most famous and successful parks in the world, we were frequently called by policy officials visiting the park to answer such questions as "What are the real impacts of the Research Triangle Park?" and "Can RTP be replicated back in ———?" While there existed many interesting anecdotal accounts of the history of RTP, its value to the region, and some of the reasons for its success, we could find no systematic evaluation of the economic development impacts of RTP or any other research park.[1] Nor did an adequate empirical base exist for conducting such a study. To be able to answer the questions posed to us with a modicum of validity, we were compelled to conduct our own study. We also became aware of the large differences among research parks in the dimensions described above and wondered which (if any) of those factors or conditions could explain the variation in the success of research parks as stimulants for technology-led regional economic development. In short, we sought to explain the "critical success factors" behind research parks.

The success of the Research Triangle and Stanford research parks and the economic booms in the San Jose–San Francisco area, Route 128 in Massachusetts, the Research Triangle, and the Austin–San Antonio corridor in the 1970s and early 1980s led many officials in regions whose economies were disproportionately concentrated in slow-growth or declining industries and hard hit by back-to-back recessions to attempt to emulate their success. Research parks also have proliferated as a result of the competition among branches of state universities. Legislative politics often make it easier to spread state government investments in research parks among several campuses. Finally, park managers and developers have become organized in the promotion of new research parks with the creation of the Association of University-Related Research Parks (AURRP) in 1985.

The result of these forces has been an explosion of research park development. The 116 parks that currently exist, and that we include in our analysis, represent a small proportion of all parks that have been started, and dozens more are now in the predevelopment stage. The title of a popular journal article, "Growing the Next Silicon Valley," [2] symbolizes this wave of imitation and the search for the holy grail of the next Apple Computer Company.

For us, this seemingly hasty adoption of the research park idea raised some troubling questions. In general, we became concerned about the huge amount of investment by state and municipal governments in the building of research parks. We wondered whether the expectations of economic development officials and others were realistic or whether they were falling into a "high-tech trap." More specifically, we saw several potential problems with the proliferation of research parks: (1) the supply of research parks seemed to be outpacing the demand for them, (2) conditions in many regions did not seem conducive to the location and growth of R&D activity, (3) the opportunity costs of research park development seemed high since other, more cost-effective strategies for restructuring a region's faltering economy could be pursued, and (4) there seemed to be a false perception that all population and economic groups shared in the net economic benefits that R&D-based development generates in a region (or "a rising tide lifts all ships"). We undertook this project to address these problems and, consequently, to shed some light on the efficiency, effectiveness, and equity of public investments in research parks. In that regard, we were motivated by John Friedmann, who said: "Planners who would interfere in regional development must understand the process by which it is generated." [3]

When conceived, this study had the following two major research objectives:

1. To assess the impact of research parks on regional economic development, including job creation, new business formation, and average wage and salary levels.

2. To assess how the benefits of such parks are distributed among population groups, particularly among minorities and women.

Parks on the scale of the Research Triangle Park in North Carolina, Stanford Research Park in California, and a few other parks in North America might be expected to have major impacts in their regions that extend beyond the economic development consequences defined above. Strong cases can be made that these parks have led to changes in the social fabric, political culture, governance, and land use patterns in their respective regions. There is also little doubt that parks such as these have significant economic development and other impacts outside their immediate regions. As noted above, the perceived success of early parks, notably the Stanford and Research Triangle parks, have contributed to the explosion of research park development in the United States as well as in Western Europe and the Far East.

By limiting our study of research parks' impacts to regional economic development, we are not saying that these other consequences are not important. Rather, our desire to conduct a comprehensive evaluation of research parks had to be tempered by the limits of our resources and by our assessment of where we could make a significant contribution to the policy literature.

This book is organized into nine chapters. Chapter 1 provides an overview of the population of research parks in the United States and describes the scope of the study and the methodology used for estimating economic development impacts. Chapter 2 discusses the expected economic development outcomes of research parks based on a review of the relevant regional development literature. In Chapter 3 we introduce a model of the stages of development of research parks and discuss the concepts of success and failure in terms of those stages. Chapter 4 presents the results of a cross-sectional analysis of a large sample of research parks. We focus on the systematic factors or conditions that help to explain the relative success or failure of research parks and that provide a basis to generalize the results. Chapters 5–7 contain case studies of three of the most successful research parks in the United States—the Research Triangle Park, the University of Utah Research Park, and the Stanford Research Park. In those chapters we attempt to identify the contextual and situation-specific factors of success that highly detailed data from multiple sources allow. Chapter 8 explores the two-way relationship between research parks and their associated universities. First, we review evidence about the importance of universities for the success of research parks; then, we look at the effect research parks have on universities. Finally, in Chapter 9 we discuss the implications of our findings for the design of regional economic development and technology policies.

Acknowledgments

While conducting this study, we were fortunate to receive invaluable assistance from many sources. First, we acknowledge the two organizations that provided financial assistance to us, the Ford Foundation and the Forum for College Financing Alternatives (a unit of the National Center for Postsecondary Governance and Finance). In particular, we wish to thank David Arnold of the foundation and Richard Anderson of the forum for their support of this project. We also must acknowledge the Association of University-Related Research Parks and its executive director, Chris Boettcher, for making membership information available to us, helping us make contact with the park directors, and endorsing our mail survey.

Individuals at the three case study parks and associated universities were generous with their time and knowledge. Robert Leak and John T. Caldwell, former executive director and interim executive director, respectively, of the Research Triangle Foundation, and Elizabeth Aycock, the longtime secretary of the foundation, submitted to lengthy interviews. William F. Little, University Distinguished Professor of Chemistry at the University of North Carolina and an architect of the Research Triangle Park, clearly appreciated the importance of our task and provided us with rich accounts of the park's history and detailed comments on early chapter drafts.

At the University of Utah, we received warm hospitality, general assistance, and detailed information on the park's history from Charles Evans, the research park director. James T. Brophy, the university's vice-president for research, and Anthony W. Morgan, the vice-president for budget and planning, provided valuable comments and insights about the university's relationship with the research park. Jan Crispin, from the University of Utah's Bureau of Business and Economic Research, served ably as our local contact in Salt Lake City.

Zera Murphy and Kirt Pruyn of Stanford University's Department of Lands Management were most hospitable during our field visit. They provided information and office support. George Naugles, a Stanford engineering graduate, served well as our local contact. Henry Lowood, the history of science librarian in Stanford's Department of Physics and a noted chronicler of the park himself, identified pertinent bibliographic information and provided helpful comments on the chapter draft. We are particularly grateful to Alf Brandin, one of the founding fathers of the Stanford Research Park and former vice-president of the university, for submitting to approximately four hours of questions in his home and by telephone.

During our field visits we also interviewed dozens of business representatives, university officials, and appointed and elected officials. We thank them for their candor and good humor. We also thank the seventy-six park managers and hundreds of business executives who responded to our mail questionnaires and telephone calls.

A long list of colleagues helped by offering advice and reading draft chapters. Ken Erickson, Ron Ferguson, Helen Ladd, Rick McGahey, Ed Malecki, Ann Markusen, Barry Moriarty, and John Rees served on an external advisory board for the project that met in Chapel Hill in June 1988 to review our study design and early progress. Ed Bergman, David Dill, and Emil Malizia, all faculty members at the University of North Carolina, took an active interest in the project and were available for suggestions and ideas when we needed some fresh thoughts. Colleagues at the Vienna University of Economics, especially Professor Walter Stöhr, also provided helpful comments on earlier drafts of the manuscript. We both visited Vienna during the past two years and made presentations on the work in progress.

Many students in the Department of City and Regional Planning at the University of North Carolina served ably as research assistants. Judith Barnet, Dawn Donaldson, Sara Flaks, Rosalind Kotz, Marge Victor, and Suk-Chan Ko all contributed to different parts of the study by collecting, organizing, or analyzing the voluminous data that have been used.

Finally, we wish to thank our editor, Paul Betz, for making the publication phase of the project relatively painless for us.

Despite all this help, we may well have erred by omitting key facts and events, or by drawing incorrect inferences from our data. We alone are to blame for these shortcomings. We also assume sole responsibility for the views that are expressed in this book; those views do not necessarily reflect the position of the Ford Foundation or any of the other organizations that assisted us.

Technology
in the Garden

1 Introduction

This chapter provides an overview of the population of U.S. research parks, offers an operational definition of research parks, and describes the principal methods that were used in the analysis that follows. Those methods include the procedure used to select case study sites, the strategy employed to collect primary data, the design of the cross-sectional analysis in Chapter 4, and the procedures used to estimate the regional economic impacts reported in Chapters 5–7.

An Overview of the Population of Research Parks

The earliest research parks in the United States were created in the 1950s. Today, only four of the 1950s-vintage parks remain in existence. Though we do not know the exact number of parks *created* between 1950 and 1989, the period covered by this study, we do know that research park formation has been cyclical and, since the late 1970s, has increased at an exponential rate.

Research parks are like Schliemann's Troy: each additional vintage is layered on top of the surviving parks of earlier vintages. We see today some 116 research parks, albeit in different stages of development and maturity (see Appendix A for an enumeration). The population of research parks is highly fluid, not only because of their high birth and mortality rates, but also because the very definition of the term *research park* is somewhat arbitrary. In the context of research parks, *mortality* refers to a park either that has failed as a real estate or business venture and ceases to exist at all, or that has changed from a research park to an office, industrial, or mixed-use park. Franco and others have reported that about one-half of all announced research parks never achieve viability and one-half of those that do are forced to diversify from research to other types of functions.[1] Figure 1-1 shows the vintage distribution of the population of parks

Figure 1-1. *Age Distribution of Research Parks* (N = 116)

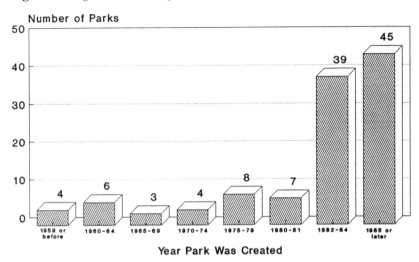

Year Park Was Created

that exists today. Not surprisingly, there was an explosion of research parks in the 1980s, particularly in the second half of the decade.

Research parks vary considerably in their location and immediate economic environment, organizational/legal status, size, and management policies. The parks that currently exist in the United States are distributed across all U.S. Census regions and among forty-two states (see Figure 1-2). They are located in urban areas of all sizes, ranging from the largest metropolitan areas to small cities hundreds of miles from the nearest metropolitan area. Some parks are situated in old, rehabilitated factory or warehouse buildings in dense parts of central cities, while others are laid out along winding roads in low-density, green, campuslike, suburban environments.

The organizational attributes of parks vary as well. Nearly 25 percent are units of public or private universities. Another 16 percent are owned by state or municipal governments. Twenty-three percent are nonprofit corporations or foundations. Fifteen percent of existing parks are owned by for-profit corporations, while the remaining 21 percent are joint public-private ventures.

The size of research parks, measured in aggregate employment, ranges from 0 to 32,000 (see Figure 1-3). The average research park has a workforce of about 1,700 employees, but a handful of very large parks sharply skews the size distribution. In fact, the large majority of research parks have workforces of less

Figure 1-2. *Geographic Distribution of Research Parks*

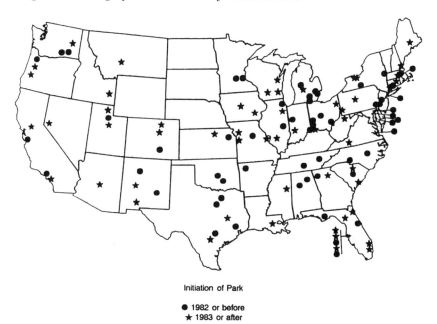

Initiation of Park

● 1982 or before
★ 1983 or after

than 200. As shown in the figure, more than thirty parks have no employment at all; they exist on paper only.

Finally, research parks differ in the development strategies and policies that park managers adopt. Those differences, in turn, reflect differences in the particular objectives of the "investors," in the perceived economic development needs of the region, and in the types of experience and styles of park management. For instance, many parks target the R&D branch plants of multilocational corporations, while others focus on generating start-ups with local entrepreneurs and nurturing small, innovative-oriented business start-ups. The nature of the physical facilities—for example, the overall land use density in the park and the existence of multitenant buildings and incubators—and the types of services provided by the park management often reflect the targeting strategy chosen.

Figure 1-3. *Size Distribution of Research Parks* (N = 112)

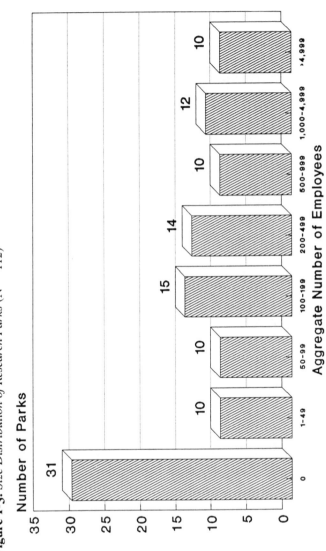

Number of Parks

35
30
25
20
15
10
5
0

31   10   10   15   14   10   12   10

0   1-49   50-99   100-199   200-499   500-999   1,000-4,999   >4,999

Aggregate Number of Employees

1987 Data, from Survey by Authors

Definition of Research Parks

Research parks (alternatively called science or technology parks) are defined here as organizational entities that sell or lease spatially contiguous land and/ or buildings to businesses or other organizations whose principal activities are basic or applied research or development of new products or processes. This definition excludes entire high-tech centers or corridors, such as Route 128 in Massachusetts or Silicon Valley in California, where high-technology businesses have concentrated outside of formal organizations. It also excludes industrial parks, in which manufacturing is the primary focus, and office parks, where administration or sales are the main functions. Business incubators that provide space in multitenant buildings for new, small businesses may be included under this definition if those businesses are R&D-oriented.

We should state that there is no "pure" science or research park. Parks often include some service-oriented businesses—for example, hotels, restaurants, banks, law offices, accounting firms, and child care centers—as well as some businesses primarily engaged in production rather than R&D. We collected estimates from park managers on the percentage of scientists and engineers employed in the park and on the percentage of organizations in the park in which R&D is a principal function. We included a park in our study population (and dubbed it a "research park") if a majority of its organizations had research and/or development as the principal function. For parks that did not yet have tenants, we considered the stated policy of the park with regard to permitted uses and the intended proportion of R&D-based organizations when deciding whether or not to include the park in our study population. When the evidence was ambiguous, we tended to be inclusive.

Principal Study Methods

We have employed two types of research designs, or methods, in this study. The first is the multiple case study, where the particular research park and its surrounding region is the case or unit of analysis. Here, *region* means the commuting area of the park. In two of the cases (the Research Triangle Park in North Carolina and the University of Utah Research Park in Salt Lake City), the region is coincident with the metropolitan statistical area (MSA) (as defined in the *1980 Census of Population*). The third case (the Stanford Research Park in

California) is a two-county area that includes a single-county metropolitan area (the San Jose–Santa Clara MSA) and one county (San Mateo) from another metropolitan area.

The advantages of the case study method are the abilities to incorporate current and historical contextual factors—political, social, and personal—directly into the analysis and interpretation of the results, and to collect highly detailed primary data from a variety of actors and textual sources. Among the disadvantages of the case study method is the high cost per case. When resources are limited, the researcher often is required to limit the number of cases. With a small sample size, it is difficult to generalize the case study results to the full population (research methodologists refer to this problem as *low external validity*).

The second method used in this study is a quasi-experimental design employing nonequivalent control groups and multivariate statistical analysis. According to this approach, we attempted to collect data from both primary and secondary sources for the full population (116) of research parks and their respective areas. Seventy-two of these responded to our primary data requests and formed our sample for the cross-sectional analysis. To establish the control groups, we also collected secondary data for a sample of areas that had no research parks but otherwise were "similar" to the areas with parks. We used multivariate statistical models to compare and explain economic growth and performance differentials between areas with and without parks to estimate the net average impact on area economic growth of having a research park, controlling for other relevant factors.

The quasi-experimental design method had the advantage of providing a large number of observations of areas both with and without parks, since the cost per unit of observation was low. It was the large number of observations that allowed us to control effectively for rival factors (besides the existence of a research park) that may have affected an area's economic growth performance. The large number of observations also allowed us to generalize to the full population of parks to ensure *high external validity*.

One disadvantage of the quasi-experimental design proved to be the difficulty in identifying a control group that was similar to areas with research parks in all important ways. In practice, we could not control for all other rival factors in a formal sense. Instead, it was necessary to rely on common sense and experience in particular cases to rule out some factors. A second disadvantage was that in order to have standardized measures for all the cases, some valuable information about some parks became inadmissible, including much of the contextual and historical factors that may have contributed to a particular park's success or failure. For these reasons, we used a combination of several case studies and a large-sample statistical analysis with quasi-experimentation. This hybrid approach allowed us to maximize the internal and external validity of the results, given the resources available.

CASE STUDY SELECTION

We selected three parks for the case studies. Since it was not possible to choose a representative sample, we decided to study three relatively mature parks that were generally believed to be successful. Given that, we sought diversity in location, regional economic structure, and university affiliation. In the cases of mature parks, enough time would have elapsed for us to observe any outcomes resulting from the research parks' establishment and development. By using successful parks, we could attempt to ascertain the critical success factors that, if discovered, would be instructive for future public policy and private investment decisions. This is not to say that the experience of unsuccessful parks would not be instructive. Rather, it is more difficult to obtain reliable information on unsuccessful parks from key informants.

After discussing the selection criteria with colleagues who were familiar with research parks, and after communicating with park managers about their interest and willingness to cooperate in a case study, we finally selected the Research Triangle Park in North Carolina, the University of Utah Research Park in Salt Lake City, and the Stanford Research Park in California.

PRIMARY DATA COLLECTION

We obtained information from a number of "key actors" at each of the three case sites. The information was collected using a closed-end questionnaire, sometimes administered in face-to-face interviews and sometimes by mail. Separate questionnaires were prepared for (1) the park manager or director, (2) the population of businesses and organizations located in each park, (3) a sample of high-technology businesses located outside the park but within the designated region, (4) key administrators in universities affiliated with the park, and (5) selected state and local government officials and business leaders. A copy of the questionnaire administered to park managers is included as Appendix B. The questionnaires for the park manager, CEOs of the largest in-park organizations, most university administrators, and local government and business leaders were implemented in face-to-face interviews. Questionnaires for the remaining actors were sent in the mail.

For the Research Triangle and Utah cases, the sample of businesses located in the region but outside the park was taken from a list of organizations in selected high-technology industry sectors obtained from the respective state employment security agency's unemployment insurance data file (the ES-202 establishment file). The industry sectors, which are shown in Table 1-1, comprise selected three- and four-digit industries from the following 1972 standard industrial classification (SIC) codes: 28 (chemicals), 29 (petroleum products), 34 (fabricated metal products), 35 (machinery), 36 (electrical and electronic equipment), 37 (transportation), 38 (instruments), 73 (business services), and 891/892 (miscellaneous professional services). These industries include those sectors generally classified in the literature as *high-technology* and *advanced services*.[2] We then excluded all establishments that first appeared on the file prior to the year in which the respective research park was created, since we were interested in whether, and how, the existence of the research park was a factor in the location of the business in the region. The resulting pared files contained approximately 1,400 and 1,500 businesses for the Research Triangle and Salt Lake City areas, respectively, and served as the sampling frames.

The California Employment Security Commission declined to make a list of establishments from San Jose and San Mateo counties available to us on the grounds that it would breach the confidentiality of those businesses. The best available alternative source of California industrial establishments was a commercial vendor. We obtained the corporate names, names of CEOs, addresses, and employment sizes for some 1,550 businesses located in Santa Clara and San Mateo counties that belonged to the industrial sectors listed above.

We drew an approximate one-in-three sample of establishments, stratified by SIC category and county, to yield a sample size of 500 for each case site. Questionnaires were sent by mail to each establishment in our three samples. Our initial round of responses was lower than desired from the California and Utah samples, so we sent a second mailing from a second, smaller, stratified random sample. One additional round of follow-up notices was sent to all nonrespondents from the first and second samples. The final numbers of completed, usable responses from out-of-park businesses were 148 for the Research

Table I-I. *High-Technology Industry Sectors*

SIC	SECTOR
281	Industrial inorganic chemicals
282	Plastics and synthetic resins
283	Drugs
285	Paints and varnishes
286	Industrial organic chemicals
287	Agricultural chemicals
289	Miscellaneous chemicals
291	Petroleum refining
348	Ordinance
351	Engines and turbines
353	Construction equipment
356	General industrial machinery
357	Office computing machines
361	Electrical transmission equipment
362	Electrical industrial apparatus
365	Radio and TV receiving equipment
366	Communications equipment
367	Electronic components
371	Motor vehicles and equipment
372	Aircraft and parts
376	Space vehicles and guided missiles
381	Engineering, laboratory, scientific instruments
382	Measuring and controlling instruments
383	Optical instruments and lenses
384	Medical instruments and supplies
386	Photographic equipment and supplies
387	Watches, clocks, and parts
737	Computer programming services
7391	Commercial R&D laboratories
7397	Commercial testing laboratories
891	Engineering and architectural services
892	Nonprofit research agencies

Triangle area, 104 for the Salt Lake City area, and 89 for San Mateo–Santa Clara counties.

CROSS-SECTIONAL ANALYSIS

The cross-sectional analysis allowed us to examine the variation in park "inputs" among the population of U.S. parks, and to begin to relate the variation in inputs to variations in regional economic development outcomes. Prior to this analysis, the same questionnaires that were administered to the park managers or directors at the three case study parks were mailed to the set of managers or directors of the full population of 116 research parks in the United States. This list of 116 parks was winnowed down from a larger list (approximately 160) of parks that (1) had been referred to as research parks in relevant trade journals (for example, *Business Facilities*, *High Technology*, and *Site Selection Handbook*), (2) had been mentioned by park managers with whom we were in contact, or (3) were listed as members on the rolls of the Association of University-Related Research Parks. To ascertain whether the majority of the parks' resident organizations engaged in research or development as their principal function, which would qualify them as a research park, and to identify parks that had gone out of business, we obtained information on the 160-some parks from research park brochures, from secondary sources, and from telephone calls to particular park offices. We retained on our final list of 116 parks those that formally existed as organizations and met these criteria even though they did not yet have any tenants or employees.

Information sought on park "inputs" included characteristics of park management; the variety of services and amenities provided to park occupants by park management, the university, or state/local government; the types of planning and land use restrictions imposed by park management and municipal planners; the revenue sources and financing techniques employed; locational attributes; the extent and nature of university affiliation; and general regional economic conditions. (See Appendix B.)

Regional economic development outcomes were measured in two ways. First, we examined the responses of park managers/directors to the series of questions regarding their perception of how the area would have differed if the particular park did not exist, all else being equal. Second, using our quasi-experimental design, we compared the area employment growth rates of each county with a research park to the rates of a comparison, or control, group of counties without parks. The control group consisted of counties belonging to the same size class of region and in the same U.S. Census division as each area with a park. There are four size classes (defined using 1980 data): nonmetropolitan area, metropolitan area with a population less than 500,000, metro area with a population between 500,000 and 999,999, and metro area with a population of 1,000,000 or more.

Average annual employment growth rates were estimated for the county with a park and its respective control group of counties for the following periods: t − 3 through t and t + 1 through t + 4, where t is the year when the particular park was established, for those parks in the population that were created in 1984 or earlier. The t − 3 through t period gave us a pretest measure of the difference in employment growth rates between areas with parks and their respective control groups. The t + 1 through t + 4 interval gave us a posttest measure that controlled for variation among parks in their degree of maturation.[3]

Following standard analysis procedures for an interrupted time series with a nonequivalent control group design, we compared the "gain" in the average annual employment growth rate (posttest minus pretest) for each area with a research park to the gain in the average annual employment growth rate for the respective control group of areas.[4] The analysis, described in detail in Chapter 4, consisted first of determining whether there were statistically significant differences between the gains in employment growth rates of areas with parks and those of their respective control groups. Second, using several alternative multivariate statistical models, we attempted to explain the variation in the differences in gain (between each area with a park and its respective control group) among the research parks in terms of various park input factors and regional economic conditions.[5]

ESTIMATING ECONOMIC IMPACTS
IN THE CASE STUDIES

We used the primary data from the completed questionnaires and published secondary data to estimate the impact of each of the three research parks on business growth, employment growth, personal income levels, income inequality, the labor market condition of women and minorities, other local labor conditions, innovative capacity, and political/civic culture of the respective region. The method and steps used to estimate all but the growth of businesses and employment were straightforward and do not require any special explanation.

The estimation of the number of businesses and jobs that research park development has "induced" in a given region consisted of the following steps:

1. Estimate the number of organizations and jobs inside the park that would not exist in the region if the park had not been established. This estimation was based on the response of each organization to the question: "Would you have located in this region if a suitable site had been available but there were no research park?" We attached a probability of locating in the region if there had been no park to each possible response: very likely = 0.9, likely = 0.7, maybe = 0.5, unlikely = 0.0, and very unlikely = 0.0. The probabilities were summed over all respondents and then inflated to the full population of R&D organizations in the park to obtain an estimate of the number of organizations inside the park whose location in the region could be attributed to the park. The amount of employment represented by these organizations was estimated by multiplying the respective probability value by the amount of employment in the organization and then summing and inflating to the full population.

2. Estimate the number of high-technology businesses and jobs outside the park that would not exist if the park had not been developed. This estimation was done in the same way as for organizations inside the park, using questionnaire responses from the out-of-park businesses that had located in the region after the park had opened.

3. Estimate the amount of employment generated in the region as a result of input purchases in the region by the organizations and businesses estimated to have been induced to the region by the research park in steps (1) and (2). This estimation was based on the questionnaire responses from in-park and out-of-park businesses on the annual value and spatial distribution of their nonlabor purchases. We used the estimated amount of local purchases to estimate the increment in regional payroll by applying a labor coefficient (the percentage of total business inputs paid to labor) from an input-output table for each region.[6] We then converted the payroll figure to an employment estimate by using average annual salary figures for workers in each of the industry sectors supplying inputs to park and out-of-park high-tech businesses.

4. Estimate the regional employment gain from the income multiplier. This was calculated for each case study area by multiplying the sum of the increments to regional payroll in steps (1), (2), and (3) by the em-

ployment multiplier in the household column of the respective regional input-output table. The employment multiplier indicated the number of jobs created in the region in, say, the retail, consumer services, housing construction, and utilities sectors for each $1 million of earnings of households living in the region.

5. Sum the estimated employment from steps (1), (2), (3), and (4) to get the estimated number of jobs that can be attributed to the research park. There might have been a small amount of double counting in (2) and (3) since some of the high-tech businesses induced to the region because of the park actually trade with park businesses. On the other hand, we used only direct purchases in (3) rather than applying a multiplier in order to offset possible double counting, as well as to remain on the conservative side in the estimation process.

Our estimates are admittedly rough, since they are based on organizations' responses to counterfactual questions, and hence may not be fully reliable. We believe, however, that they provide an order of magnitude of research parks' employment effects and can serve as useful indicators in an overall policy assessment.

2 Research Parks and Regional Economic Development

Theoretical Expectations

The decision to build a research park often precedes careful evaluation, as is frequently the case when economic development initiatives originate at the state and local level. Most existing studies of research parks are anecdotal, describing park characteristics rather than analyzing outcomes.[1] When the success or failure of parks has been studied, it has been as real estate ventures alone.[2] Existing studies do not address some key analytic questions that, if answered, would help guide planners and policymakers in deciding whether a prospective research park is likely to be an effective tool in changing the structure and performance of the regional economy. These questions include (1) what types of regional economic development outcomes can we expect from research parks in general? and (2) how might those outcomes vary, and what set of factors might account for the expected variation in outcomes?

In this chapter, we provide answers to these questions by applying the relevant literature on regional development to generate hypotheses about the expected outcomes of research parks. We critically review those hypotheses and provide some insights about their applicability in the context of the United States.

Regional Development Theories

To facilitate discussion, we divide the standard theories of regional development into two groups: (1) those that stress diffusion, of growth or innovation, for example, from a center outward, and (2) those that focus on amenities of location that enhance a region's "creativity" or dynamism without an explicit spatial transmission process.[3] The first group includes growth pole/growth center and innovation diffusion theories in which relationships among businesses or individuals are of primary importance. These explanations are based on an initial imbalance within the region at large. Policies consistent with these theories normally target key (propulsive) industries or innovative individuals.

The second group of regional development theories includes entrepreneurship, seedbed, and regional creativity theories in which environmental factors and nonspatial synergies contribute to the area's overall "fertility." These explanations do not necessarily begin with intraregional or sectoral imbalances. Policies consistent with these theories focus generally on individuals or places rather than industries.

We can divide diffusion or "growth transmission" explanations of regional development into three groups: growth pole/growth center theory, hierarchical diffusion theory, and nonhierarchical diffusion theory. Growth pole/growth center theory, originally formulated by François Perroux, states that investments in propulsive industries (the pole) in strategically located centers induce growth by firms in technologically related industries through the formation of backward and forward linkages with the propulsive industries. The "center" is defined as the locus of innovation. In Perroux's original statement of the theory, the induced growth occurred in an "economic region" that did not have to be spatially contiguous to where the propulsive industry was located. In later revisions of the theory, induced growth occurred within a geographic region and emanated from a growth center.[4]

Like growth pole/growth center theory, hierarchical diffusion theory stresses the spatial filtering or trickling down of innovations. But, whereas the revisions of growth pole/growth center theory focus on the transmission of growth from the center to hinterland within a region, hierarchical diffusion theory specifies a downward filtering through a system of cities and region. Specifically, innovations are transmitted from larger to smaller metropolitan areas. Hierarchical diffusion theory also differs from growth pole/growth center theory in its emphasis on process innovations. Moreover, hierarchical diffusion theory has been used to explain the spread of research parks themselves as a type of organizational innovation.[5]

Some diffusion theorists have argued that empirical data on interindustry transactions do not support growth pole/growth center or hierarchical diffusion theories of innovation transmission.[6] Innovations do not diffuse in a spatially systematic way; rather, intraorganizational linkages within multiplant firms account for many of the transactions of goods and information that are observed. These linkages are typically interregional. Similarly, Higgins has argued that growth pole/growth center and hierarchical diffusion theories are no longer relevant in today's information-age economy. "Today's communications make it perfectly possible," he says, "for an innovation made in Stuttgart to reach Detroit before it reaches Bonn." [7]

The differences among these theories have important implications for which level of government should develop research parks. If the research park, as the center of growth, induces further growth in the region, either adjacent to the park or in a surrounding hinterland, park development might be an appropriate state/regional/local strategy. On the other hand, if the bulk of economic benefits are realized outside the region, a higher level of government (multistate regional or central government) should be responsible for park development. [8]

Of the three variants of diffusion theory, growth pole/growth center doctrine provides the most extensive foundation for explaining research park development. The parks can be viewed as growth centers that lead to the development of localization and agglomeration economies, specifically high concentrations of R&D activity and amenities that are attractive to well-paid scientists and engineers. [9]

The propulsive industries—the sectoral targets of investment in the case of research parks—require some discussion. Applied to the research park strategy, the appropriate target is a particular function, namely research and development, rather than a particular industry sector, as is assumed in growth center theory. Thus, investment in R&D in old-line industries may be as appropriate in some regions as R&D in high-tech industries. On the other hand, it could be argued that research universities constitute the analogous propulsive industry or growth pole. Investments then would be targeted not only to research universities directly, but also to those enterprises most likely to take advantage of the resources associated with research universities.

The types of induced economic growth predicted by growth pole/growth center doctrine are (1) existing firm expansion and new business formation through backward and forward linkages, (2) new business formation in the same industry(ies) as the propulsive industry through localization and agglomeration economies, and (3) firm expansion and new business formation in consumer services and retail trade from the indirect and induced growth in residential-based activities. The bulk of the expected induced growth would be through backward and then forward linkages, since the propulsive industries would be chosen in growth pole/growth center doctrine so as to maximize the interindustry trade multiplier effect.

The types of expected induced economic growth from research parks, according to growth pole/center doctrine, would be weighted more toward new business formation through the development of localization economies, including spin-offs, and growth from residentially based trade and services. The former would be due to the highly competitive conditions associated with industries in the early phases of their respective product cycles and the external economies of sharing a highly specialized labor force.[10] The latter would stem from the relatively high wages and salaries in R&D activities compared to manufacturing. A related, more indirect source of growth stimulation can occur if the research park helped increase the research productivity of universities or other local research institutions, which in turn could induce the location of new firms that would find association with those institutions attractive.[11]

The effectiveness of research parks, according to growth pole/growth center doctrine, rests on the assumption that a trickling down (spread) process dominates a polarization (backwash) process. *Backwash* works against the development of the surrounding (peripheral) region, especially in an intranational context, in that the initial investments at the center draw the most productive labor and possibly other mobile factor inputs from the periphery in search of better opportunities. *Spread* works for the intended development of the periphery through the center's firms purchasing their inputs from firms in the periphery, through center firms' new investments in the peripheral region, and through the center's absorption of the periphery's surplus labor, leading to increased per capita incomes in the periphery. This, in turn, can stimulate further growth in the periphery through the normal consumption multiplier.

Decisions about the types of investments to be made in the center, and in which centers research parks should be located, are critical, but growth pole/growth center theory does not provide sufficient guidance. Generally speaking, the doctrine implies that the designated centers should be those that offer the greatest agglomeration economies. Similarly, we infer that the investments would be targeted to those technology areas (for example, microelectronics, biotechnology, or machine tools) that potentially offer the greatest multiplier effects and competitive advantages in their respective regions compared with the rest of the world. The trickle-down effect also works best in the absence of trade barriers, such as tariffs or quotas, which normally are absent in an intranational context.

Another issue relates to exogenous versus indigenous development strategies as applied to research parks. Exogenous development refers to that which
is initiated, propelled, and controlled by organizations located outside the region. Indigenous development refers to development that is regionally initiated
and planned. It places emphasis on small- and medium-sized regionally owned
enterprises, rather than on recruited branch plants of large multilocation corporations. The provision of information transfer and advisory services for local
propulsive businesses, as opposed to financial incentives for nonlocal firms, for
example, would be another attribute of an indigenous strategy.[12]

In principle, there is no reason why growth center strategy could not follow
an indigenous development path so long as the target region had the requisite
resources to start with. There is an empirical and practical question of whether
initial investments in an indigenous development strategy would be sufficiently
large to generate threshold levels of demand to create forward linkages and
agglomeration economies. Some research parks in the United States are based
on the recruitment of R&D branch plants of large multinational corporations
(for example, the Research Triangle Park and the Princeton Forrestal Center);
others are based mostly on the development of new, locally owned businesses
that are linked to nearby research universities (for example, the University of
Utah Research Park, Atlanta's Advanced Technology Development Center, and
the New Haven [Connecticut] Science Park).

The indigenous/exogenous distinction is important for another class of theories, grouped here on the basis of their common focus on the fertility of the
region as a propagator of new economic activity. The intent of the policies
based on these theories is to foster self-generated growth. For proponents of
these theories, the role of research parks is to serve as a seedbed or focus of
creativity for the region.

Entrepreneurship/seedbed/creativity theorists ask: Why have some locations
(such as Austin, Route 128, and Silicon Valley) been economically more dynamic than others, and what role can public policy play in creating and fostering those conditions?[13] Andersson provides a general answer: "Creativity as a
social phenomenon," he says, is "primarily developed in regions characterized
by high levels of competence, many fields of academic and cultural activity,
excellent possibilities for internal and external communications, widely shared
perceptions of unsatisfied needs," and synergies among local actors. He and
others suggest that these conditions can be influenced by public policy, including the creation of research parks. These conditions are also related to city

size, since larger cities typically provide a wider variety of services and greater agglomeration economies than small cities do.[14]

Modern applications of the entrepreneurship/seedbed/creativity theories relate the density of entrepreneurs and extensiveness of entrepreneurial networks to rates of growth. In this regard, "[t]he major economic and social benefit of a Science Park is achieved when the environment fosters new business development by innovative entrepreneurs." In particular, these applications stress an "entrepreneurial phase" of park development that follows an "institutional phase." During the institutional phase the parks attract major research facilities, add services and support operations, and develop a critical mass of scientists and innovators who begin to interact. Then, during the entrepreneurial phase, scientists and engineers, singly or in teams, spin off new enterprises, usually located in the same region. Policies to increase the density of entrepreneurs focus either on the potential innovators by providing technical assistance, peer support, specialized training, or start-up capital, for example, or on the overall cultural and economic environment.[15]

Since research parks, by definition, are characterized by a concentration of highly skilled workers, agglomeration economies, and large amounts of R&D, they are a seemingly obvious tool with which to create an innovative environment. The parks that are developed in accordance with entrepreneurship theory differ somewhat from those based on diffusion theory: they contain incubator space for start-up entrepreneurs, as opposed to permanent facilities for established firms or new branch plants. We already have noted several examples in the United States—Atlanta's Advanced Technology Development Center, the New Haven Science Park, and the University of Utah Research Park. Generally, however, parks with incubator facilities are more prevalent in Europe than in the United States. These parks typically operate by providing incubator space for a fixed period of three to five years until the start-up firms are able to move into permanent facilities elsewhere. The provision of low-rent space in close proximity to essential services and other entrepreneurs greatly reduces the mortality rate of new firms—from over 90 percent outside of incubators to approximately 50 percent within them.[16] In some places, spin-off parks develop adjacent to the parks with incubators (for example, the Weitzman Institute in Israel, the Plassey Technology Park in England, and the Chalmers Park in Sweden).

For proponents of entrepreneurship theory, two questions remain outstanding: whether a sufficient pool of local entrepreneurs exists or can be developed, and whether local economic conditions can support new start-ups. In some locations, public intervention may be necessary to overcome the inevitable start-up problems that new firms experience at enormous costs. In these places it may be better to attract branch plants of established firms, at least in an initial stage during which essential infrastructure can be developed. This approach seems to characterize the development of the Research Triangle Park. During the park's early years (the 1960s), the Raleigh-Durham economy was not well developed. The primary industries—tobacco, textiles, and furniture—were in stagnation and increasingly controlled from outside the region. There was virtually no local venture capital, poor transportation and communication networks, and little history of successful entrepreneurship in science- or technology-based industry sectors.

Another critical question for policymakers to ask is whether successful start-ups will generate sufficient local multipliers in the long run to justify local support. If they are linked primarily with firms outside the region, there will be considerable leakage of potential economic benefit. In that case, as noted above, higher levels of government would be more appropriate sponsors of the parks.

Research Parks and Expected Regional Development Outcomes

By applying the theories reviewed above and bringing to bear the results of empirical studies of R&D location, spin-offs, and technology diffusion, we can outline a set of expected regional development outcomes of research parks. Factors that may affect the magnitude and spatial incidence of the expected impacts, including location and regional structure, park attributes, and external linkages to businesses, universities, and other research institutions in the region, can then be discussed.

Figure 2-1 classifies the possible impacts of research parks on regional development. Primary impacts are defined as those that result in changes in the magnitude of economic activity—for example, the number of businesses and jobs, personal income, and value added. These primary impacts have some relevant distributional dimensions: spatial, sectoral, occupational, socioeconomic, and so forth. Secondary, or derivative, impacts are those that derive from the primary changes but result in changes in the economic structure, macroeconomic

Figure 2-1. *Types and Dimensions of Potential Park Impacts on Regional Economic Development*

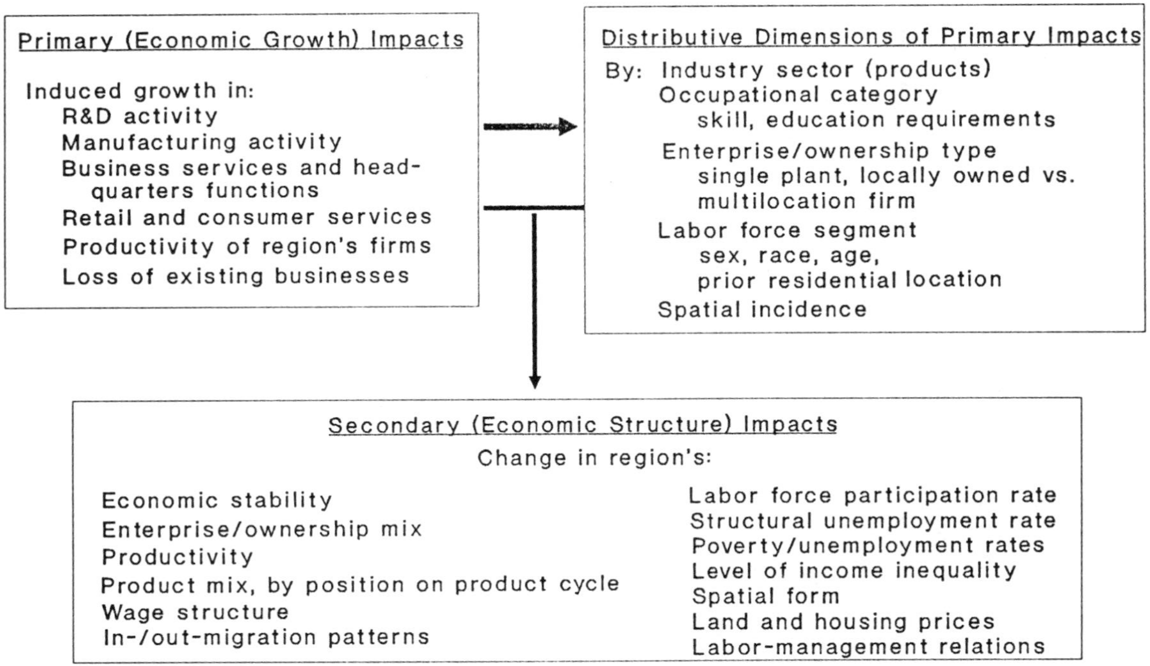

Primary (Economic Growth) Impacts

Induced growth in:
R&D activity
Manufacturing activity
Business services and head-
    quarters functions
Retail and consumer services
Productivity of region's firms
Loss of existing businesses

Distributive Dimensions of Primary Impacts

By: Industry sector (products)
    Occupational category
        skill, education requirements
    Enterprise/ownership type
        single plant, locally owned vs.
        multilocation firm
    Labor force segment
        sex, race, age,
        prior residential location
    Spatial incidence

Secondary (Economic Structure) Impacts

Change in region's:

Economic stability   Labor force participation rate
Enterprise/ownership mix   Structural unemployment rate
Productivity   Poverty/unemployment rates
Product mix, by position on product cycle   Level of income inequality
Wage structure   Spatial form
In-/out-migration patterns   Land and housing prices
                                            Labor-management relations

performance, and spatial organization of the affected region as a whole. We focus here on the primary impacts, particularly on the generation of economic growth (existing establishment growth and new business formation) through the mechanisms of backward and forward linkages, localization economies, intra-corporate organizational linkages, and the earnings multiplier (see Table 2-1 for a summary). Of the secondary impacts, we consider only the spatial incidence of induced growth.

The most sizable development impact within the regions in which parks are situated is likely to be an induced growth of R&D activity. R&D businesses are more likely than other types of businesses to cluster within the region in order to take advantage of the sharing of a specialized labor force, facilities and expertise in external research institutions, business services, particular types of social and cultural milieus, and access to technical and market information through proximity to R&D facilities in competing firms.[17] Consequently, once a region "takes off" with a successful research/science park, it should continue to experience growth in the R&D sector.

Generally, the induced growth of manufacturing and other activities through backward linkages will be smaller from an R&D complex than from a manufacturing complex, for at least two reasons. First, R&D activity is heavily labor-intensive. Material inputs represent a relatively small portion of the total inputs to R&D facilities. Second, firms within industries that typically produce intermediate manufactured goods for R&D activities (for instance, instruments, computers, and machine tools) tend to be highly spatially diffused themselves and often serve national markets.[18] Also, to the extent that a large proportion of R&D facilities are branches of large, multilocational, vertically integrated corporations, the potential demand for regionally produced inputs will be less than otherwise expected. This is because many of the inputs will be obtained from other corporate branches located outside the region. For these reasons, it is expected that backward linkages from research parks are less a source of induced growth than growth center theory by itself predicts.

Induced growth through forward linkages can result when manufacturing facilities that utilize the output of the R&D establishments locate in the region to gain access to the sources of technological innovation. The expected magnitude depends on how scientific and technological innovations diffuse, and are adopted, over space. Rees reports that adoption rates of new process technologies in manufacturing are, in part, a function of proximity to centers of innovation.[19]

On the other hand, advances in the rapid communication of technological innovations originating from universities and independent research institutions may make physical proximity of manufacturing activities to external R&D activities less important. Spatial proximity of R&D activity to manufacturing is probably most important for multilocational manufacturing firms with large investments in R&D and in industries with rapid technological change (or short product cycles). In these cases, corporations are more likely to locate their manufacturing plants close to their own R&D facilities.[20] Yet, on the whole, significant induced growth through forward linkages is not expected because of the spatially diffused, national and international markets for R&D.[21]

Induced regional economic growth is expected to result from the increase in local consumer demand that comes from the increased population and earnings growth associated with additional employment in new and expanded businesses linked directly or indirectly to the park. The magnitude of this should depend mostly on the magnitude of the total induced growth through localization economies and backward/forward linkages since, for a given size class of region, the earnings multiplier would not be expected to vary much across locations.

What are the principal factors that may explain the variation in the magnitude and spatial incidence of the impacts of research parks? We discuss three types of factors: (1) the economic structure of the center and peripheral region, (2) the internal organization and management of the park, and (3) the external linkages between the park and key institutions, including state and local government, universities and other research institutions, and the region's existing set of businesses.

ECONOMIC STRUCTURE

Growth center doctrine emphasizes that investments should be made in centers where external economies (agglomeration, localization, and urbanization) are the highest.[22] However, recent empirical research in the United States suggests that localization economies (that is, the location of other R&D activity) are the most important of these externalities.

Specifically, the research shows that (1) R&D is highly clustered in a relatively small number of areas, (2) a suitable supply of scientists, engineers, and technicians is one of the most important factors for the spatial clustering of R&D, (3) agglomeration economies seem to matter to several R&D industries,

Table 2-1. *Potential Primary Impacts of Research Parks on Regional Economic Development*

Type of Impact	Immediate Source of Impact	Mechanism	Comments
Location of new R&D activity	Park enterprises, university, other R&D activity, milieu	Localization economies	Growth will depend on amount of R&D in the region, strength of region's universities in tech–related areas, and/or presence of government research labs.
R&D firm spin–offs	R&D enterprises in park; scientific faculty brought to region	Localization economies	The rate of spin–off activity varies by enterprise ownership, type of R&D activity in and out of park, and university regulation of faculty entrepreneurship.
Location of manufacturing activity	R&D enterprises in park and induced R&D activity	Backward linkages	Material factor inputs are a small fraction of total R&D costs; leakage from region is typically high but varies by type of enterprise and degree of vertical integration.
Business services location	R&D enterprises in park; induced R&D, manufacturing and other functions	Backward linkages	Depends on enterprise ownership, types of R&D firms, and any induced manufacturing firms.

Intrafirm manufacturing location	R&D enterprises in park; induced R&D in region	Forward linkages	Depends on the importance and frequency of face-to-face contact between R&D and manufacturing functions within the firm and on the corporate organization of R&D.
Location of other intrafirm functions	R&D enterprises in park; induced R&D and manufacturing in region	Forward linkages	Depends on enterprise ownership, type of R&D and manufacturing activities, proximity of R&D and HQ functions, and supply of skilled labor. Large metro most likely to attract HQ and sales functions.
Retail and consumer services growth	New households from induced in-migration to region's labor force	Earnings multiplier	Magnitude depends on total amount of induced growth (first 6 types of impacts) and the new workforce's pay level. Minimum leakage from region.
Generalized new business development	All sources listed above	Urbanization economies	The larger the region, the higher the magnitude. Amenities and quality of public management may be important.
Increased productivity of existing firms	R&D activity by park enterprises and in region's university	Technology transfer	Depends on match between R&D and technology needs of region's industries. Innovation adoption rates vary between existing and new firms and by effectiveness of marketing services to region's firms.
Loss of business (industrial gentrification)	Park enterprises and induced enterprises with high pay	Area wage roll-out	Depends on magnitude of wage/salary differences between existing and new firms, and ability to transfer labor skills among existing firms.

including biotechnology, microelectronics, and communications, (4) metropolitan area size, as an indicator of agglomeration economies, is highly correlated with the concentration of private, industrial R&D facilities, but not with the concentration of other institutional forms of R&D (for example, universities and government laboratories), (5) areas with federal research facilities and research-oriented universities, which are disproportionately located in smaller areas, are associated with high concentrations of R&D employment, and (6) areas with high concentrations of manufacturing activity are associated with high concentrations of R&D activity. Others have argued that the quality of local air service (measured, for example, as the number of flights per day) is an important factor in the location of R&D facilities.[23] And some researchers have observed that multilocational, technology-oriented firms prefer to locate R&D facilities in the same area as their headquarters.[24]

All of this suggests that two types of local economies are likely to support park development and to generate appreciable induced growth of other R&D activities in the region. Large agglomerations (metropolitan areas) can provide the breadth and depth of labor supply, business services (including air transportation), proximity to headquarters functions, and, in the northeastern and north-central parts of the United States, proximity to large concentrations of manufacturing activity. What they may lack is proximity to the top research universities (and large federal research facilities) and residentially based amenities that managers, scientists, and engineers indicate are important in their locational preferences.

Smaller metropolitan areas with either a major research university or a prominent government research facility as a magnet (or "anchor") are also likely to be conducive locations for research parks. These areas often initially lack the depth of business services, headquarters functions, and prior concentration in manufacturing. In addition, they may lack depth in certain specialized labor skills—for example, those of technicians. Some of this business service infrastructure may develop later, as in the case of the Research Triangle area. Less diversification of the industrial structure and less depth in the number of research institutions (compared to the largest agglomerations) imply that research park establishments and induced R&D activity might be more specialized in particular application sectors (for example, microelectronics, pharmaceuticals, biotechnology, and robotics), depending on the strongest research departments in the universities, the areas of application in any federal research facilities that might be located in the region, or any sectoral specialization of existing private R&D establishments.

To the extent that induced growth through backward linkages occurs at all, its magnitude should vary by the depth and breadth of the relevant manufacturing sectors in the region. Everything else being equal, the location of research parks in larger agglomerations that have a specialized manufacturing base in scientific instruments, machinery, or electrical and electronic equipment should induce more existing business expansions and new plant start-ups through backward linkages. Induced growth in the region through forward linkages from a research park is most likely to occur in large metropolitan regions if the relative costs of manufacturing production workers are not too high compared to smaller centers and peripheral regions. Relatively underdeveloped regions surrounding smaller centers with concentrations of R&D activity would also be likely candidates for induced growth through forward linkages if the demand for requisite labor skills is not too specialized. Otherwise, firms would not locate in such areas because of the nonavailability of the necessary labor pool.

One potential negative growth impact whose incidence may vary by regional economic structure is *industrial gentrification*. This withering of the traditional industrial base is likely to occur if the wage and salary levels associated with the research park and its induced growth are significantly higher than those of the region's traditional industries, and if many of the labor skills are transferable from the traditional to the new industries. Under these circumstances, businesses in the region's traditional industries may not be able to attract and keep a labor force due to upward wage pressure and thus may be forced out of business in that location. Industrial gentrification is more likely to occur in smaller, more isolated labor markets with labor forces that lack depth and diversification than in large metropolitan regions.

PARK ATTRIBUTES

The internal characteristics of research parks are also expected to affect the regional development impacts. These characteristics include the park's recruiting policy and resulting mix of organization types, the existence of government facilities as anchors, restrictive covenants, the existence of incubator facilities within the park, and the ease of interaction among park enterprises. A park's recruiting policy often stipulates whether park management seeks the R&D

branch plants of large multilocational firms (as in the Research Triangle Park), for instance, or attracts or develops new small- or medium-sized firms (as in the University of Utah Research Park), or cultivates a mixture of these (for example, the Stanford Research Park). We suggest that the incidence of spin-offs from park enterprises is higher when the park has a larger proportion of relatively new, small- and medium-sized firms, and that the induced growth of other firms outside the park through both forward and backward linkages may be higher as well.

The choice of recruitment policy will affect the type of physical facilities available to park tenants as well as land/facility pricing (or leasing) policies.[25] Newer and smaller firms will require smaller structures, smaller sites, lower site development costs, and perhaps greater interaction with other park enterprises for specialized inputs, including technical and market information. Thus, at one extreme, a research park can be designed to serve as an incubator of R&D enterprises. On the other hand, parks that seek only the R&D branch plants of large corporations will use covenants to restrict activities (such as manu-facturing), to set minimum-sized sites and floor areas, and to control physical appearance and landscaping. These restrictions increase the cost of locating in the park, often making it prohibitively expensive for new and small enterprises.

The ease of interaction among park organizations is a function of the physical layout of a park, the extent of planned events such as colloquia and seminars, and the internal policies of park tenants. The tenants in a research park might prefer minimal, or tightly controlled, opportunities for external interaction in order to minimize labor force raiding and the leaking of innovations or strate-gies to competitors. In any event, when interaction among park organizations is low, we can expect the incidence of spin-offs to be low as well.

Finally, a park's ability to attract a large government research facility to serve as an anchor is hypothesized to be positively related both to the continued via-bility of the park itself and to the amount of induced growth in the region. That is because government research facilities, unlike the R&D branch plants of many large, vertically integrated corporations, are more likely to subcontract to regional firms and to purchase other inputs from within the region.

EXTERNAL LINKAGES

The extent and nature of linkages between research park enterprises and various types of nearby institutions, including universities, businesses, and state and local governments, are expected to contribute to the success of the park and the amount of induced economic growth in the region. Universities can provide research park enterprises with specialized laboratory facilities, faculty expertise for consulting, and a regular supply of graduates for entry-level professional jobs, as well as the prestige of association. The university provides localization economies by drawing to the region other R&D activities, which in turn may share those resources. The linkage also can benefit the university through the growth of research contracts, consulting opportunities and dual appointments as an inducement in faculty recruiting, and "wired" job opportunities as an aid in student recruitment at the graduate level.[26]

The university-park linkage is expected to create localization economies that attract new R&D activity to the region and spin-offs (for example, by faculty and private scientists/engineers forming their own firms) when (1) the relevant university departments are conducting research at the leading edge of the field, (2) the research has applications to process and product innovations, and (3) university policies support and encourage entrepreneurial arrangements between private research establishments and the institution and individual faculty members.[27]

The link between a research park and a region's existing businesses may also provide benefits in both directions. Innovation diffusion and technology transfer from park enterprises to businesses in the region can increase the latter's productivity. The likelihood of technology transfer within the region will depend on the match between the region's industry mix and the R&D activity, as well as on the type of enterprises in the region; small, independent firms are probably more likely to seek technology developed outside the region (but see note 27 above).

The park also can benefit from the activities of the region's businesses. In the early stages of park development, those businesses can help market the park to prospective tenants and help secure venture capital for them, as was the case at the Stanford Research Park.[28] In addition, the business community can affect the incidence of induced growth due to spin-offs and other new business development by providing technical and managerial assistance and assisting in the provision of venture capital. The extent to which the business community helps the park achieve viability and increase the incidence of induced growth will depend on how well established and prominent the region's businesses are, as well as on the attitudes of the business community and the park's enterprises toward each other. These attitudes, in turn, may depend on the degree of competition between the preexisting businesses and the new R&D and related organizations over scarce factor inputs, particularly labor. Regional economies that have been performing well, with low unemployment rates, might, therefore, be more likely to experience frictions. On the other hand, regional economies with chronic economic distress might be most receptive to R&D and R&D-induced economic development. Yet, ironically, the existing businesses in distressed regions may be least capable of forming linkages that will benefit the new technology-based businesses.

Almost all research parks in the United States receive some support from state or local government, although the type and degree of that support vary. First, many parks are public corporations or subsidiaries of public universities. Others are privately owned but are recipients of various kinds of government subsidies including land, buildings, services and infrastructure (for example, roads, water, and sewer facilities), and property tax reductions. Less direct government subsidies to research parks include, at the state level, the provision of specially designed economic development, education, and job training programs and, at the local level, favorable land use policies that favor expansion. Government subsidies effectively reduce the cost of doing business for both publicly and privately owned parks and thus, in general, for the parks' tenants. To the extent that some of the government subsidies, particularly infrastructure, are of a public goods nature, they will also be available to enterprises that locate outside the park but within the region. These assistance programs, which offer widespread benefits, perhaps represent the most cost-effective type of government subsidies, especially if they are strategically timed; consequently, the more they are used, the greater the expected induced impact from government intervention.

The Role of Research Parks in Regional Development and Policy

There is a growing consensus among state and local policy officials and academics that a region's long-term economic prospects will depend on its ability to generate and sustain a concentration of enterprises capable of developing new products that can penetrate national and international markets.[29] If we assume this to be the case, we might then ask: What is the appropriate role of research parks, given their prospective regional development outcomes, within the broader domain of regional innovation policies?

To provide tentative answers to this question, we first need to address the issue of whether a regional concentration of R&D activity is necessary to generate a regional concentration of enterprises with the ability to innovate. Implicitly, this also raises the issue of the importance of spatial proximity of research and development to manufacturing activity. When an R&D presence leads to additional innovative start-up businesses that, in turn, foster local dynamism, we have what Norton and Rees call a "seedbed." [30] But under what circumstances can innovative firms be present *without a concentration of* private R&D activity in the region? There are at least three plausible scenarios.

The first scenario is a regional concentration of manufacturing branch plants of large, but innovative, multilocational corporations. R&D input is supplied from other branch plants located outside the region. This is not unusual in the United States; examples include Dallas, Texas, and Phoenix, Arizona. The downside to this scenario is that the region's economy is less stable, since the control and decision centers for the innovative enterprises are exogenous to the region.

The second scenario applies where there are indigenous enterprises whose primary R&D input comes from universities, research institutions, or private R&D enterprises outside the region. While we know of no particular area that fits this description, it is possible, in principle, if spatial proximity between R&D and its application in manufacturing on an interorganizational basis is not essential for innovation diffusion and adoption to occur on a timely basis.

A third scenario appears when new small- and medium-sized, innovative-oriented manufacturing firms develop the necessary R&D capacity internally. Typically, these types of firms are the least able to invest in R&D because of the lack of capital; as a result, they experience high failure rates. But the provision of R&D grants or equity capital by state and local governments may help ameliorate this problem. In this scenario, economies of scale in research and development on a regional level are not as important since each firm designs its R&D effort to be consistent with its specific product development plans. In fact, it may be difficult even to measure R&D activity in these cases since it would physically occur within facilities devoted primarily to manufacturing.

On the other hand, if a concentration of R&D activity is deemed essential for the development of a regional complex of innovative manufacturing firms, there are at least two strategies to consider. The first is to help existing regional enterprises become more innovative through intraregional technology transfer and innovation diffusion. The source(s) of innovation could come from the establishment of university-related technology development centers, industrial extension services, or the stimulation of autonomous R&D businesses, as well as from the creation of a research park.

Both university-related technology development centers and industrial extension centers have become increasingly popular in state technology policies. University-related technology development centers, most often funded by a consortium of private corporations in the region, tend to operate on a semi-proprietary basis (for example, the Microelectronics Center of North Carolina and New York's Centers of Advanced Technology program). If operated in this way, the opportunities for innovation diffusion to smaller regional enterprises are restricted. Establishment of an industrial extension service, on the other hand, is a less costly way to provide R&D services and potentially is more accessible to smaller and less capitalized firms. It should be noted that neither of these R&D sources needs the physical or institutional form of a research park.

The stimulation of autonomous R&D enterprises in the region would be based on the assumption that existing area manufacturers would be more likely to adopt technological innovations with a greater concentration of R&D activity in the region. We argue, however, that this is not likely to occur without specific arrangements between individual R&D enterprises and individual manufacturers. The creation of a research park might be the most effective instrument for autonomous R&D activity in a region (if other regional conditions are suitable), but university-related technology development centers and industrial extension services represent promising alternatives.

The second strategy for developing a regional concentration of innovative firms through regional R&D activity is to provide incubators that could generate and nurture new, small, innovative firms from their inception. These firms would depend on external sources of R&D as well as other essential business services. Spatial proximity of a number of such firms could lead to economies in the external supply of these services, although the provision of R&D services, in particular, might not be as dependent on the spatial clustering of demand as would more general types of business services such as communications and the repair and maintenance of business equipment.

Conclusion

There seems to be little doubt that research parks *can* stimulate regional economic development. The structure of the regional economy itself will have a strong effect on regional economic development outcomes. Our evidence suggests that regions need to have a sufficient population to support agglomerations of economic activity, *or* possess a strong research university, to serve as good hosts for research parks. Yet, these locational factors may not be sufficient to ensure the success of research parks. Organizational characteristics of the enterprises that locate in the park, leadership and long-term commitment by sponsoring institutions, and external linkages that develop between park organizations and other regional institutions and businesses all can mediate the expected outcomes.

Even when it is reasonable to expect a research park to stimulate economic development in the region, there will also be some costs or disadvantages to consider. For example, few additional job opportunities may be generated for low-skilled members of the labor force. In any event, available theory leads to a prediction that research parks may be a cost-effective economic development tool and thus a viable development strategy for only a limited set of regions.

3 Defining Success and Failure in the Context of Research Park Development

Public officials, university administrators, economic development planners, and others who have expressed an interest in research park development for their regions, or who actually have helped to establish parks, have proceeded more with the *assumption* that those parks can be successful than with the demonstrated fact. This does not mean that those park advocates and developers are imprudent people. Rather, it reflects some difficult conceptual and methodological problems that have prevented careful analysis of the performance and the potential of research parks.

One of the conceptual difficulties is that there is no consensus about the definition of success. Success is a normative concept, so one must have a frame of reference—or, in this instance, a set of goals—against which to measure it. One problem is that the key actors in the park development process often have different goals in mind. The most commonly cited goals relate to economic development. But both the literature and our data from interviews with park developers, elected officials, university administrators, business leaders, and others confirm the existence of other goals, including technology transfer, land development, and enhancement of the research opportunities and capacities of affiliated universities.

Even if we accept economic development as the raison d'être of research parks (as we do for analytic purposes in this book), we confront some difficult problems. First, "economic development" is a broad moniker. To assess success fully we need to understand what aspects of economic development are particularly sought by park creators, including, for example, job generation, income growth, greater income equality, expanded opportunities for special groups within the labor force, and economic restructuring of the region. These objectives normally are not mutually consistent.

Second, the contribution of a research park to any of these aspects of economic development is hard to measure. Some park promoters cite as evidence of success the employment and payroll of park businesses. For some of the larger parks those numbers are impressive; however, they tell only part of the story. Many of the jobs located in a park may well have been generated within the region even if the park had not been created. Yet, even if we were to focus on these "induced" benefits of park development—that is, outcomes that can be attributed directly to the establishment and operation of a research park—they provide an incomplete picture of economic development impacts. One must consider, as well, induced employment and payroll in businesses *outside* the park that have located in the region, or have been able to expand, because the park was established.

Third, to measure properly the success of research parks in terms of their effect on regional economic development (or any other goal), one must consider the costs as well as the benefits of a research park strategy. Those costs include direct expenditures by government on land acquisition and infrastructure development, tax expenditures from any financial inducements used by government, and the opportunity cost of the land for research parks versus other types of uses.

Finally, we must be able to account for park failure, which, like success, is a normative concept. The challenge here is to identify stages during the park development process that are success (or failure) thresholds and particular characteristics of parks and their environment that relate to the likelihood of success or failure.

Goals and Investment Strategies

One way to measure the success of research parks is to assess their performance against stated goals, as written into legislation and found in documents and interviews. (This type of success is sometimes referred to as "policy effectiveness.") This section considers the goals park developers and other key players have given for investing in research parks, as well as the nature of their investments.

Table 3-1 ranks the objectives for park creation as understood by seventy-two park managers/directors whom we surveyed. Economic development objectives are cited most frequently, followed by objectives relating to the university

Table 3-1. *Park Managers'/Directors' Perceptions of Research Park Objectives*

IMPORTANCE OF ORIGINAL OBJECTIVE	RANKING[*] #1	#2
RELATIVELY IMPORTANT: *Economic development*		
To diversify region's economic base	2	1
To develop and nurture new businesses	1	2
To capitalize on existing R&D in region	3	3
To expand local employment opportunities	4	4
AVERAGE IMPORTANCE: *University and technology development*		
To enhance university's technical training via collaborative research	5	5
To increase technology transfer by park businesses	6	6
To encourage entrepreneurship in region	7	7
To increase region's productivity via innovation	9	8
To expand employment opportunities for local university graduates	11	9
To commercialize university–based research	10	10
To enhance prestige of affiliated university	8	11
RELATIVELY UNIMPORTANT: *Income/profit generation and redistribution*		
To provide higher–paying jobs locally	12	12
To maximize profits from park land/facility sales/leases	13	13
To expand employment opportunities for low–skilled workers	14	14

[*] Ranked in order of importance, based on two weightings of survey responses. Ranking #1 attaches weights of 0.55, 0.375, 0.225, and −0.15 to "very important," "moderately important," "somewhat important" and "not important," respectively. Ranking #2 is based on a data sort, with each response as a successive key.

Table 3-2. *Services Provided by Park Management*

SERVICES PROVIDED BY PARK MANAGEMENT	NUMBER OF PARKS	% OF RESPONDENTS (N = 69)
Liaison with universities	59	84.3
Sewer and water services	58	82.9
Road construction and maintenance	54	77.1
Landscaping and grounds maintenance	50	71.4
Signage	49	70.0
Natural gas hookups	48	68.6
Liaison with local government agencies	46	65.7
Liaison with state government agencies	45	64.3
Land use planning	44	62.9
Security services	32	45.7
Prewired telephone systems	31	44.3
Conference facilities for meetings	27	38.6
Restaurants and social facilities	22	31.4
Business services	19	27.1
Management consulting	19	27.1
On-site hotel	18	25.7

and technology development. Relatively little importance is given to the enhancement of opportunities for minorities, and, since many parks are operated as nonprofit corporations, park managers do not consider profitability to be an important objective. It is important to note that the information in Table 3-1 reflects the distribution of parks by organizational status (see Figure 4-4).

The desire to improve the environment of park firms and, consequently, to contribute to job creation, has guided the investment strategy of park management. It typically invests in infrastructure, facilities, and services for park occupants, using private funds (if the park is a for-profit corporation or a joint public/private venture), university or state and local government contributions, special assessments, and/or internally generated funds from current or projected land sales or rentals. The list of services provided by park management is presented in Table 3-2.

First, park management makes investments that relate to the immediate physical environment and facilities of individual resident organizations, including the construction of multitenant buildings that provide space for service, office, and small R&D enterprises; the provision of sewer and water service; gas hookups; ground maintenance and landscaping; signage; and land use and site planning. In many cases these investments are discretionary on the part of park management, although some types of infrastructure investments may be required by local government as development exactions in some instances.

Second, park management often provides business services for the convenience of the resident organizations. These include conference and meeting facilities, an on-site hotel, management consulting, and restaurant, recreation, and child care facilities. The provision of these services by the park is based on the concept of localization economies.

Third, park management often acts as a liaison between the occupants and other institutions in the area, including universities and state and local governments. Park managers provide these types of services because other actors have not taken the responsibility (especially in the case of infrastructure); because resident businesses cannot undertake them themselves, at least as efficiently;[1] and/or because the managers expect that the investments will make the park more attractive to organizations and, hence, increase the site rent they can charge.[2]

Other officials whom we interviewed—be they in public office, the universities, the local business and investment communities, or the enterprises located in the park—specified other goals and thus have invested in research parks for different reasons.

PRIVATE INVESTORS

Many private individuals, including some who are identified as venture capitalists,[3] provide equity to research park development companies for the purchase of land, site improvements, marketing, and other expenses.[4] These investors view research parks as a real estate venture with the potential to generate a positive net cash flow and capital appreciation. As such, they typically are less interested in the research and technology orientation of park occupants than they are with the economic viability of the park. Thus, if the research park is not sufficiently profitable, these private investors often lobby park management for a relaxation of park land use restrictions so that office, commercial, warehousing, and even manufacturing firms can locate there.[5]

STATE AND LOCAL GOVERNMENT

State and local governments provide direct and indirect subsidies to businesses and park developers with the justification that the parks will serve to stimulate regional employment growth and well-paying jobs for local university graduates.[6] This motivation recently has been strongest in stagnant and declining areas of the country.

The most prevalent (and costly) form of government investment is the provision of infrastructure to or within parks—for example, access roads, new highway interchanges, and water and sewer lines—when these are not provided (and paid for) by the park managers/developers as development exactions. In some cases, state or local governments also abate property taxes for businesses in the park (in approximately 24 percent of the parks surveyed), dedicate public lands for park use (21.1 percent),[7] lease or purchase properties in the parks for government use (19.7 percent), and construct facilities in the parks, often creating anchors (15.5 percent).[8]

UNIVERSITIES

Universities invest in research parks for two reasons. In some cases, universities act as a real estate developer and manager on land they already have in their portfolio. (The best-known example of this is the Stanford Research Park.) In other cases, universities buy land—or keep land bequeathed to them, rather than sell it—to develop as a research park. (The development of the Forrestal Center by Princeton University is one of several recent examples.) Universities do this to earn a positive rate of return, to ensure a supply of space for current and future university expansion, or to have a nearby site for the location of technology-oriented private businesses that can help faculty develop and commercialize inventions.[9] The latter motivation reflects a growing awareness within universities that there are important opportunities for private sector–university cooperation in applied research and technology development.

PARK OCCUPANTS

The last group of investments are made by the organizations that locate in research parks. Most park occupants construct their own buildings, either on land they own or on land they have leased on a long-term basis. Organizations that lease space in buildings, either from the park or a management company, often retrofit the buildings to their own specifications. Alternatively, the lessor does the retrofitting. Park occupants also invest in essential services that others do not provide for them, including water, sewer, and gas hookups; access roads; and telecommunications facilities (for example, microwave transmitters). Firms that conduct R&D or light manufacturing with processes that use or produce hazardous materials often construct their own containment facilities.

Stages of Park Development

The investments of park developers or managers, private investors, state and local governments, universities, and park occupants are not necessarily made at the same time but rather over the lifetime of a park. We can, in fact, think of research parks as having three distinct stages of development: incubation, consolidation, and maturation. Each period has different expected outcomes and therefore different criteria for measuring success.

The first event during the *incubation stage* is the conception of the idea to develop a research park. That idea can be conceived by university leaders, economic development officials in the state or local government, elected politicians, or even private developers. If the initial idea "floats," some type of feasibility study typically is undertaken next. That may be a formal study conducted by a consulting group or an informal study conducted by a local committee. Such studies usually explore the prospective ability of park management to recruit appropriate R&D organizations to the park and the availability of financial and institutional resources that would be required to build and sustain the park. As part of the feasibility assessment, consulting groups and local committees often visit existing research parks, particularly those commonly believed to be successful.

If the feasibility study indicates that the necessary resources for park development are available and the park has a good chance of attracting appropriate R&D organizations (or if local factions are intent on proceeding with the park in any case), an initial governing structure is created, the required legal documents (including a charter for the park corporation) are drafted, and a formal announcement of the park is made. In this study, we have used the formal announcement as the indicator of park "birth" and the zero point in counting the park's age. We regard parks that are conceived as an idea but not "announced" as "not born" and do not count them as failures. Some parks that are considered, but not born, are terminated during or after the feasibility study, and others have had charters and initial governing structures but die for several reasons before an official announcement has been made. To continue our analogy, we can refer to those parks as "stillborn" rather than failed.

After a new park is announced, a series of detailed planning studies, fundraising, and recruitment activities are usually undertaken. The initial governing structure may be revised. Land must be purchased or conveyed and the initial infrastructure must be created, including the construction of access roads, water and sewer mains, and electricity service. These considerable up-front expenditures make research parks a relatively risky undertaking, especially if they are made before any organizations have committed to locate in the park.

In many parks, building construction does not begin until the prospective tenant signs a lease or closes on a land purchase. The commitment by the first R&D organization is typically parlayed into a major media event by park management. In some cases, private organizations take options to buy park land but do not commence construction immediately, even though their purchase has been announced publicly. In other cases, park management builds a multi-tenant "spec" facility as a further inducement to locate in the park. The ability of park management to land a major R&D organization that can serve as an anchor early after the announcement of the park's existence seems to be one of the critical success factors in this stage of park development.

The incubation stage ends when the first R&D organization (or organizations, if several occupants locate in the park simultaneously) is operating in the park. We consider a park to be successful in this stage if it has been able to reach this point. Many parks that are born do not reach this point, however, for at least two immediate reasons: park management was unable to muster sufficient resources to put the necessary infrastructure (physical or institutional) in place, or the infrastructure was put in place, but an R&D organization could not be found to locate in the park before operating funds and/or commitments ran out or loans became due. Underlying the immediate reasons for failure in the incubation stage are a variety of situational factors. Prominent among these, based on conversations with a number of research park directors, are (1) overoptimistic feasibility studies, (2) withdrawal of support from key leaders or institutions (for example, legislators or university trustees) when positive outcomes occurred more slowly than expected, (3) unexpected changes in the macroeconomy that made borrowed funds more expensive than planned or the pace of

R&D activity slower than envisioned, and (4) poor management in terms of the formulation and implementation of the park's business plan.

Three outcomes can occur for a given park during its incubation stage:

1. The park entity goes out of business, or ceases to exist, because it did not attract any R&D organizations.
2. The park ceases to exist as a *research* park, but it is transformed into a general business or office park and achieves viability in that form.
3. The park achieves success in that it has attracted at least one R&D organization.

No matter what the outcome is for the park developers, state and local governments typically experience a net revenue loss in the incubation stage since infrastructure investments and other improvements are made before significant new tax revenues begin to accrue. In the early years, the net revenue loss is exacerbated if the state and local governments use tax abatements, reductions, or credits or nontax assistance programs (including building and land subsidies, subsidized loans, and grants) to induce firms to move into the park. The net economic development benefits to the region at this stage tend to be small, if positive at all.

In the second stage of research park development, the park land begins to fill up with new tenants. We call this the *consolidation stage*. Marketing of the park becomes the primary activity of the park's management. When successful, new jobs and additional payroll dollars are added to the region's private economy, and new tax revenues begin to flow into state and local government coffers (assuming that tax incentives were not used extensively or have expired). In addition, jobs and payroll dollars increase in the region as park employees and their families spend part of their park earnings and local governments increase their spending on public services (this is the "economic multiplier" effect). At this stage the net economic development benefit to the region depends, in large part, on the types of businesses that have moved into the park (in terms of occupational mix, stability of employment, and so forth), and whether they relocated from within or outside the region. It also depends on the structure of state-local government finance and spending patterns. Clearly, if all new tenants moved to the park from other locations in the region, the local economy would be no better off than it had been before. If tax incentives were used, the region could be worse off economically.

The degree of success in the consolidation stage can be measured as the aggregate number of new jobs or payroll dollars directly generated by the R&D establishments that have located in the park. A normalized success measure that takes into account the size of the region might be the percentage of high-tech jobs in the region that are located in the park. Failure in this second stage of park development is not a discrete measure, as in the incubation stage. Rather, it is indicated by a relatively small level of aggregate jobs or payroll directly attributable to the park.

In the third or *maturation stage*, organizations in the park develop significant backward, lateral, and forward linkages with other businesses in the region. In addition to increased economic growth, the region undergoes changes in its economic structure during this period of maturation. There is an additional multiplier process as new business start-ups and existing business expansions are stimulated. And there is industrial agglomeration, or a clustering of businesses, which integrates the regional economy and begins to dominate the local economic structure and the region's macroeconomic performance. Further growth can occur in the form of spin-offs from the earlier organizations in the park that have grown and prospered.

In this third stage, we expect to see some negative growth effects as well. Some out-of-park businesses may no longer be able to compete for resources, particularly labor and land, as wages and prices have increased. And if growth has outstripped the pace of infrastructure development, there will be increased congestion on the area's roads and perhaps local governments will need to impose development moratoria. We also expect to find corresponding changes in the political, cultural, and social structure of the region.

The measure of success in the third stage of park development should be a function of the number of new jobs or additional payroll dollars in the region indirectly generated by the organizations in the park, and of the induced change in the economic structure of the region, which is more difficult to measure. Failure in the third stage is, again, a relative concept. A park whose organizations have not established any linkages with businesses outside the park, have not led to any spin-offs, or have not induced any new enterprises to start up or relocate to the region would have to be considered a failure in the maturation stage, although it may have been highly successful in the previous consolidation stage.

We make two general points about these development stages. First, the stages are not always discrete and distinct. In particular, there may be a blurring between the consolidation and maturation stages. Second, the time that it takes for a park and a region to develop from the incubation stage to later stages will vary considerably among parks and regions. Policies of the park developer, the extent and type of government assistance, locational attractiveness, prevailing macroeconomic and investment conditions, and just plain good luck will be important factors in determining if transitions from the first to subsequent stages occur and how long they take. Based on the parks that we have analyzed, the incubation stage typically lasts from twenty-four to thirty-six months, while a park may not reach the maturation stage for ten years or longer.

Measuring Success

As noted above, measuring success in the incubation stage of park development is relatively straightforward: research parks that are able to recruit at least one R&D organization and do not relax requirements for location (for example, that the park occupant must engage primarily in R&D) can be judged to be successful as policy initiatives. In reality, however, success in the first stage is only "potential" success because the policy effectiveness of research parks that were established with the intent of fostering economic development cannot be ascertained until at least the consolidation stage of park development. Thus, real estate viability without relaxation of location requirements is a necessary but insufficient condition for policy success.

We measure success in the consolidation and maturation stages of park development in ways suggested by the previous discussion. In our three case studies (Chapters 5–7), we measure success in terms of the number of jobs represented by the R&D organizations that have located in the park and in terms of the following *induced* changes in the region: employment growth, business start-ups, regional income and income equality, employment opportunities for women and minorities, occupational mix and the local wage structure (related to "regional restructuring"), research capacity of the local university(ies), and the business climate and political culture.

The measurement of all these induced outcomes is based on responses to survey questions about the perceived effects of the research park. For instance, we asked executives in businesses inside and near the park to judge the likelihood that their organizations would have located in the region, or exist at all, if the park had not been created. We then employed regional multiplier analysis to estimate the amount of induced enterprise and income growth within the park and region. To obtain another perspective, we asked park business and university officials to judge the effects of park creation on a university's research capacity and productivity. These approaches are described more fully in the case study chapters.

Our conclusions about the success or failure of the three case study parks are based on both objective information about costs and induced benefits and subjective judgments made from a large amount of background data that we collected in our field visits. For those three parks, we focus on the extent of new business formation and induced backward, forward, and lateral linkages, and consequent regional restructuring, which should be evident in the maturation stage of development.

Our conclusions about the success or failure of the large sample of parks, reported in Chapter 4, are based on objective data only, using the quasi-experimental research design described earlier. Because the forty-five parks in our sample are of mixed ages, we are not specific about the sources of induced growth that we observed (for example, as arising from consumer spending in the consolidation stage, or from agglomeration economies and other linkages in the maturation stage).

In Chapter 4 we limit our analysis to one type of outcome: direct and induced regional employment gains. We focus on employment changes for two reasons: (1) all actors whom we interviewed stated that the "expansion of local employment opportunities" was one of the most important goals of research parks, and (2) time-series data on employment, by county, are readily available in published sources.

The specific measure of success employed in Chapter 4 is

$$\left[\frac{(E^p_{t+m} - E^p_{t+1})/E^p_{t+1}}{(m - 1)} - \frac{(E^p_{t-1} - E^p_{t-n})/E^p_{t-n}}{(n - 1)} \right]$$

$$- \left[\frac{(E^c_{t+m} - E^c_{t+1})/E^c_{t+1}}{(m - 1)} - \frac{(E^c_{t-1} - E^c_{t-n})/E^c_{t-n}}{(n - 1)} \right]$$

$$= \text{DIFFERENCE}$$

where

E = the average annual employment

p = a county containing a research park

c = the set of control counties for each respective county with a research park t= the year that the referent park was established

m and n = forward and backward lags, in years, respectively

The terms in brackets are average annual employment growth rates. *DIFFER-ENCE* is either the cardinal difference in park/nonpark county growth rate differentials or a 0/1 index, with 0 indicating failure and 1 denoting success. We used two criteria for success versus failure. In the more lenient criterion, we assigned 1 to those parks whose *after* minus *before* growth rate difference was greater than the *after* minus *before* difference for control group counties. In the more stringent cutoff, we assigned 1 to those parks whose *after* minus *before* growth rate difference was greater than 120 percent of the *after* minus *before* difference for control group counties. We refer to the former as *MSR 100* and to the latter as *MSR 120*.

By matching control group counties to park counties on the basis of metropolitan status, size, and region, we attempted to control for selection differences between the two groups. The success measure we employed (the equation above) controlled for those selection differences by taking into account differences in prepark conditions between park and nonpark counties. Yet it would still be possible for park counties' observed growth rate increases in the years following the establishment of the parks to be due to factors that were unrelated to the research parks. An enhanced "control" would define the control groups more narrowly than we do here, accounting for industrial base composition and park location within a metropolitan area (core or periphery), as well as metropolitan status, size, and region.

The measure of success shown in the equation keeps constant the number of years subsequent to the date of park creation, during which outcomes can occur (that is, the lead period, $m - 1$, is four or five years for all parks). A variant of the equation shown above is to allow the lead period to equal $[1988 - (t + 1)]$, where t is the year of park creation, thus allowing the maximum amount of time for outcomes to become evident in each park. For the oldest parks, then, the lead period would be more than thirty years, whereas for the youngest parks, the lead period would be less than five years. This approach makes comparisons across parks spurious but still allows us to judge whether given parks succeed or fail relative to their control county(ies) as long as the outcomes for the control county(ies) are measured for the same time period $[1988 - (t + 1)]$. We report results in Chapter 4 using the above equation; we plan to use the variant in future analyses.

Conditions for Research Park Feasibility

In the previous sections of this chapter we have said that research park success depends on whether the *goals* established for the park are being met and whether the *means* used to achieve those goals are appropriate. We have argued, as well, that in assessing the success, at a given point in time, of parks that are of mixed ages, one must consider the park's stage of development. We conclude the chapter by noting that the likelihood of success, however it is defined, depends critically on locational factors, as well as on park inputs (or investments), timing, leadership, and plain good luck.

The locational attributes that are important for success include the relative spatial proximity of the region to national and international centers, or nodes of economic activity. But they also include the regional economic milieu, which has the potential to provide significant positive external economies to the businesses in the park.

The literature indicates that many regions with high concentrations of R&D activity share these characteristics: the presence of research universities and R&D laboratories, the depth and breadth of business services, a large concentration of manufacturing activity, the presence of headquarters functions in the same corporation, quality airline service, and residential amenities.[10] These factors also tend to be attributes of large metropolitan regions. Therefore, one could conclude that research parks located within large metropolitan areas have a greater chance of success than parks sited in nonmetropolitan or small metropolitan locations. That interpretation is flawed, however, because it implies unidirectional causality. In fact, headquarters location, residential amenities, and airline service may follow, rather than precede, the creation of research parks and related economic activity. We test the relationship between these locational factors and success in Chapters 4–7.

To the extent that the hypotheses advanced above—about (1) the importance of localization economies and (2) the external economies that can be provided by major research universities—hold, they suggest certain implications concerning the location of successful research parks. These implications, each of which is addressed in the succeeding chapters, are as follows:

1. Research parks located in regions that already have a concentration of R&D activity are most likely to be successful in the incubation and consolidation stages.

2. A research park by itself is unlikely to serve as a seedbed for stimulating a concentration of R&D activity in a region. Such a park is not as likely to be successful in the incubation and consolidation stages.

3. A major research university or a large, federal government–supported laboratory potentially can provide a region with the seedbed for stimulating a concentration of R&D activity. In fact, the external economies provided by these institutions in even small- and medium-sized centers and nonmetropolitan regions are probably much more important for a science/research park in the incubation and consolidation stages than those provided by large metropolitan centers without a concentration of R&D activity.

4. Those regions in greatest distress that potentially could benefit the most from the presence of a viable research park (that is, through manufacturing job creation via forward linkages, productivity enhancement via technology transfer and diffusion, and overall wage increases in the local labor market) also are where research parks are not likely to be feasible. These include older manufacturing regions and nonmetropolitan areas *without* an existing R&D concentration.

This last point deserves considerable attention because research parks have been mentioned increasingly as a possible strategy to target economically lagging regions in the United States, as well as in Europe. To the extent that market forces (rather than public policies) have the strongest effect on economic outcomes, research parks are not likely to reverse the fortunes of regional economies whose industries already have shown distress. The economic development of a state's most distressed regions will more likely be advanced through other technology and innovation policy instruments than by creating a decentralized system of research parks affiliated with, say, many of the branches of the state university system.

4
Determinants of Success and Failure

Cross-Sectional Analysis

Research parks are not all alike. They differ in size, ownership, location, services offered, types of restrictions used, appearance, relationship with universities, and other important characteristics. This chapter relates the differences in park characteristics to observed differences in regional outcomes that can be attributed to research parks.

We examine these relationships by analyzing data collected from 72 of the 116 research parks that we identified to be in existence as of January 1989. (The full list of parks is in Appendix A.) These 72 parks contain most of the research park employment in the United States at this time.[1]

Characteristics, Location, and Operating Policies of U.S. Research Parks

We often envision research "parks" as broad expanses of green space interrupted by cleanly designed low-rise buildings along curving roads in a campus environment. Indeed, the most prominent U.S. parks, including those at the Research Triangle in North Carolina, the University of Utah, and Stanford University, have those characteristics and may be partially responsible for the popular image. There are, however, a variety of physical park configurations, including relatively small, inner-city developments that contain multistory, converted factory or warehouse buildings (for example, the New Haven Science Park in Connecticut). Parks differ in many other ways, including their ownership and organizational structure, imposition of restrictions, history of land assembly, use of government subsidies, and primary industry focus.

As noted earlier, the development of research parks is a recent phenomenon. Indeed, most existing parks were opened in the 1980s. This is reflected in Figure 4-1, which shows the distribution of employment, by size class for

Figure 4-1. *Employment Growth in Research Parks*

Employment Range

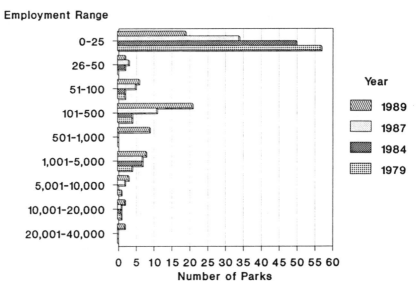

four different years, for our sample of seventy-two parks. The figure shows that fewer small parks (with 0–25 employees) existed in each successive period, and more parks entered the middle ranges over time. Still, in 1989 most parks were small; fifty-eight of seventy-two parks had no more than a thousand employees.

The distribution of park sizes among our sample is shown in Table 4-1, but in terms of acreage rather than employment. The average-sized park is built on about 600 acres, with just over 300 acres improved with infrastructure and slightly more than 100 acres developed with buildings. One park has buildings on more than 1,000 acres, but forty-five have less than 51 acres with buildings on them. On the average, there is almost 75 percent more unbuilt, improved land than there is built land and more than twice as much unbuilt, unimproved land. This indicates that many parks are underbuilt and have ample room to expand. (See Figure 4-2.)

The lands on which research parks have been developed were owned previously by public and private universities (37 percent), federal, state, or local governments (26 percent), private individuals (25 percent), real estate developers (7 percent), and others (5 percent) (see Figure 4-3). These properties were acquired with state/local government funds (in twenty-four parks), private investor funds (in twenty-three parks), university funds (in nineteen parks), and

Table 4-1. *Distribution of Parks by Size*

Total Acres	No. of Parks	Improved Acres	No. of Parks	Built Acres	No. of Parks
0–50	7	0–50	23	0–50	45
51–100	15	51–100	17	51–100	9
100–250	16	101–250	14	101–250	4
251–500	17	251–500	6	251–500	3
500–1,000	7	501–1,000	4	501–1,000	3
1,001–2,500	6	1,001–2,500	4	1,001–2,500	1
2,501–5,000	2	2,501–5,000	2	2,501–5,000	0
5,001–9,999	2	5,001–9,999	0	5,001–9,999	0
		No response	2	No response	7
Average = 594 acres	N = 72	Average = 313.5 acres	N = 70	Average = 114 acres	N = 65

Figure 4-2. *Status of Land Use in Research Parks*

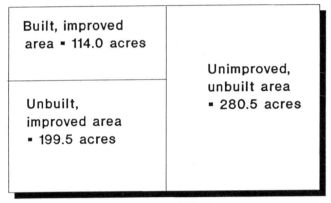

| Built, improved area ▪ 114.0 acres | Unimproved, unbuilt area ▪ 280.5 acres |
| Unbuilt, improved area ▪ 199.5 acres | |

Total Area = 594.0 acres

Figure 4-3. *Ownership of Land prior to Creation of Research Parks*

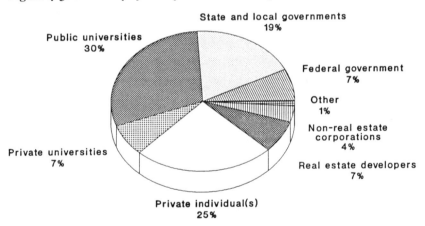

State and local governments 19%

Public universities 30%

Federal government 7%

Other 1%

Non-real estate corporations 4%

Private universities 7%

Real estate developers 7%

Private individual(s) 25%

nonuniversity private donations (in eighteen parks).[2] Venture capital has been used for equity by businesses in approximately 35 percent of the parks and by approximately 10 percent of the park developers/management corporations.

Twenty-five percent of U.S. research parks are units of public or private universities (see Figure 4-4). Another 60 percent have a formal or informal affiliation with nearby research or doctoral-granting universities even though they are not owned by them. In most of the remaining 15 percent of parks that do not have an institutional affiliation with a university, there are interactions

Figure 4-4. *Organizational Status of Research Parks*

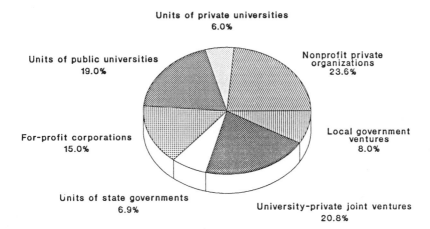

Units of private universities
6.0%

Units of public universities
19.0%

Nonprofit private organizations
23.6%

For-profit corporations
15.0%

Local government ventures
8.0%

Units of state governments
6.9%

University-private joint ventures
20.8%

between employees in their resident organizations and faculty members of a nearby university. The seventy-five percent of the parks that are not units of a university are divided among five organizational types: university–private sector joint ventures (20.8 percent of the total sample), for-profit corporations (approximately 15.0 percent), units of state government (6.9 percent), local government ventures (8.0 percent), and nonprofit private organizations (23.6 percent). Over one-half of all research parks have some government affiliation, either through a public university or as a unit of a state or municipality.

In 53 percent of the parks we surveyed, lots and facilities are sold; elsewhere they are rented. In 1989, the average value per acre was $122,250, determined by estimating the fair market price (in almost 60.0 percent of all parks), by establishing a full cost recovery price (in 19.4 percent), or by setting prices to achieve a desired rate of return (in 11.1 percent). The value of research park land, like the value of business sites in general, varies significantly from place to place. Five parks in our sample (in McLean, Va., Philadelphia, Tampa, Dallas, and Princeton, N.J.) command more than $250,000 per acre. At the other extreme, land in a park in upstate Michigan sold in 1988 for approximately $10,000 per acre. Of the land in parks that is leased, the average price in 1988 was $6.56 per square foot.[3]

The current pattern of land ownership in our sample of U.S. research parks is shown in Figure 4-5. The park developer/management corporations themselves own 30 percent of park land, affiliated universities own 38 percent,

Figure 4-5. *Current Land Ownership in Research Parks*

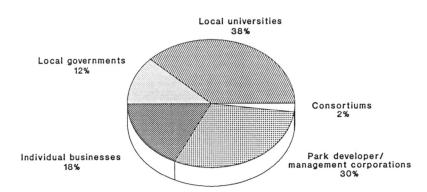

individual businesses own 18 percent, local governments own 12 percent, and public-private consortiums own 2 percent.

The organizational status and ownership of research parks are significant, in part, because of their relationship to the parks' stated goals (see Table 3-1). In our survey, goals related to economic development were listed as important by the greatest number of parks, especially parks that were owned and operated by state and local governments. The second most frequently cited group of goals related to (1) the university and (2) technology development. Parks stressing those objectives tended to be owned and operated by universities. Finally, the least frequently mentioned goal was profit maximization, which was of primary importance to the 15 percent of parks that were run by for-profit corporations.

Most research parks use planning controls to regulate the use of land or buildings. This practice distinguishes parks from other possible business sites in the same region. The use of deed restrictions (in 80.6 percent of the parks surveyed) preserves the character of the park by outlawing certain types of activity—mostly heavy manufacturing, residential, and warehousing.[4] Of the parks surveyed, 70.8 percent also regulate the "footprint," or the ratio of the building's ground floor area to the lot size, and 45.8 percent impose a minimum lot size requirement.[5] A small number (9.7 percent) require a minimum building area. These restrictions are shown in Table 4-2.

Such limitations often make the total cost of locating in a park substantial. As a result, many businesses, particularly new start-ups, are priced out of research parks unless they have access to venture capital or business incubators.

Table 4-2. *The Use of Deed Restrictions in Research Parks*

	NO. OF PARKS	%
Deed restrictions used (any type)	58	80.6
Maximum building/lot ratio	51	70.8
Minimum lot size*	33	45.8
Minimum building area**	7	9.7

* Average minimum = 2.83 acres.
** Average minimum = 30,000 square feet.

As noted above, relatively few businesses and park managers have access to venture capital. The use of incubators in U.S. parks is limited as well. Only 36 percent of park respondents (twenty-six out of seventy-two) indicated that they currently provide incubator space for small, start-up organizations. In those parks, the median percentage of floor space devoted to incubators is less than 10 percent. Another 16.7 percent of research parks (twelve out of seventy-two) planned to add incubator space soon. These relatively low percentages are in contrast to non-U.S. science parks, which place a higher priority on new technology development and incubation.[6]

Whether or not certain kinds of activities are proscribed, park management often shapes the character of the research park by selectively recruiting tenants. Table 4-3 ranks the types of organizations favored in park marketing efforts. Some parks specialize in the types of technology-oriented and R&D businesses they recruit in order to establish a critical mass within an industrial subsector and to create the greatest synergies with the local university. For example, the Research Triangle Park has concentrations of biotechnology, pharmaceutical, and electronics companies, and the University of Utah Research Park has a relatively large proportion of biotechnology and medical technology companies.

Finally, the large majority of research parks (85 percent) are located in metropolitan regions. According to 1985 U.S. Census estimates, 45 percent of these metropolitan parks (or 38 percent of the total) are in regions with over 1 million population. Parks are located in all geographic regions of the country, although

Table 4-3. *Favored Organizations in Park Marketing*

	RANKING*	
TYPE OF BUSINESS	#1	#2
Technology–related	1	1
R&D	2	2
Locally owned	4	3
Start–ups	3	4
Branches of U.S. corporations	5	5
Branches of foreign corporations	7	6
Light manufacturing	6	7

* Ranking #1 is based on a weighting of survey responses with "strong preference" for and against given a value of 0.6 and –0.6, respectively, and "mild preference" for and against given 0.4 and –0.4, respectively. Ranking #2 is based on the frequency of parks citing "strong preference for."

the South is somewhat overrepresented and the West is somewhat underrepresented. Parks that have been created since 1982 are more likely to be located in either the largest metropolitan centers or in nonmetropolitan areas.

The Impact of Research Parks on Regional Employment: Evidence

Over forty years ago, Howard W. Odum, a sociologist and regionalist at the University of North Carolina at Chapel Hill, was one of the first to hypothesize that scientific research activity can stimulate the economic development of a region.[7] Odum had the South, and specifically North Carolina, in mind, and his ideas heavily influenced the series of decisions that eventually led to the creation of the Research Triangle Park. Over time, economic development officials nationwide and in other parts of the world have come to treat Odum's hypothesis as an axiom. After visiting the Research Triangle Park, the University of Utah Research Park, the Stanford Research Park, and a small number of other established parks, many of those officials returned home to set up their own research parks; hence, the dramatic growth in the research park population. Unfortunately, park developers too often fail to account for unique local conditions or even to assess whether any park can be justified in a benefit-cost sense.

The principal questions that we now address are whether, and under what conditions, Odum's hypothesis can be sustained. Specifically, do research parks achieve their intended regional economic development outcomes? These questions are not easy to answer because (1) it is difficult to isolate the effect of research parks from other factors, and (2) we cannot observe the counterfactual case—that is, what would have happened in the putative absence of the research park. In view of these methodological difficulties, we present several different types of evidence about the regional economic development impacts of research parks—from direct questionnaire responses and from quasi-experimental and econometric analyses of data—to help answer this question.

By asking park managers how several types of economic outcomes would have differed if a research park had not been created in their region, we addressed the counterfactual question: Have research parks induced more growth in your region than would have occurred without them? In general, a majority of respondents believed that their parks had *no* significant effect on improving regional economic performance. However, of all possible outcomes, managers identified employment growth and business start-ups as the ones most likely to have been affected by the creation of parks. Of the total respondents, 13.1 percent said that their area's rate of new business start-ups would have been much lower if the park had not been established. Over 60 percent of the respondents said that the area's employment growth rate would have been much or somewhat lower.

While these responses are suggestive, their subjective basis raises some questions about reliability and validity; park managers are not disinterested observers. For that reason, we also used a quasi-experimental research design to answer the counterfactual question about park-induced economic development outcomes.

Quasi-experimental design refers to a class of research designs that apply some aspects of classical experiments—such as the use of control groups and pretest and posttest observations—to causal research in actual field settings. They differ from classical experiments because the random selection of cases into experimental and control groups is usually not a feasible (or ethical) option in field research, nor is the physical isolation of cases from all other putative causal influences. Quasi-experimental designs potentially can increase the

Table 4-4. *Direct and Indirect Impacts of Research Parks on Regional Economic Development to Date*

Impact ("If the park had not been created, how would the following have differed?")	% of Respondents (N = 62)				
	MUCH LOWER	SOMEWHAT LOWER	ABOUT THE SAME	SOMEWHAT HIGHER	MUCH HIGHER
New business start-ups would have been:	13.0	31.5	38.9	11.1	5.6
Efficiency of land use in the area would have been:	11.8	31.4	41.2	11.8	3.9
The employment growth rate would have been:	9.4	50.9	20.8	18.9	0
The overall quality of life would have been:	7.5	34.0	41.5	13.2	3.8
Importance of manufacturing to area's economy would have been:	5.7	22.6	50.9	18.9	1.9
Nonprofessionals' wages would have been:	5.7	18.9	67.9	5.7	1.9
Professionals' salaries would have been:	3.8	32.1	52.8	11.3	0
The quality of the natural environment would have been:	3.7	22.2	63.0	11.1	0
Stature of area's universities would have been:	1.9	35.8	39.6	15.1	7.5
Quality of area's universities would have been:	1.9	34.0	45.3	15.1	3.8
Area's population would have been:	1.9	32.1	58.5	7.5	0
Minorities' job opportunities would have been:	1.9	22.6	62.3	11.3	1.9
The cost of living would have been:	0	13.2	83.0	3.8	0

internal validity of research findings toward the level of validity that can be obtained in classical experimentation.[8]

In this application of a quasi-experimental design, the region's total employment growth rate served as a proxy for the rate of economic development. The use of employment growth as the principal outcome measure was justified, in part, by the results in Table 4-4. Other actors who were interviewed, including elected officials and business leaders, also cited employment growth as the most important goal for research park development.

After experimenting with different definitions of economic development success or failure (see Chapter 3 for a fuller discussion of alternative measures), we chose to use, as the measure of success, the difference in total employment growth rates—both after and before a park had been established—between counties with a research park and a control group of counties (without a park) having the same metropolitan status, population size, and location as the counties containing the research parks.[9] By matching control group counties to each county with a park on the basis of metropolitan status, size, and region, we attempted to control for "selection" differences between areas with and without research parks. Admittedly, it is still possible that increases in the park counties' growth rates following the establishment of the parks were due to additional factors not related to the presence or absence of a research park. But the chance of that has been reduced by taking into account preexisting differences between park and nonpark counties. Table 4-5 lists the success indicators of forty-five research parks that were created before 1985.[10]

As indicated in the table, the *after* total employment growth rates for park counties range from more than 10.0 percent to approximately −6.0 percent. Those numbers mean little unless they are compared to some benchmark, so the control group counties' growth rates are also shown. The respective *difference* in growth rates ranges from +9.25 to −9.75 percentage points. Thirty-two of the forty-five parks are in counties that grew faster than their control group counties in the years after the parks were established. Using a more stringent success criterion, which requires park counties to grow 20 percent faster than control group counties, we find that twenty-six of the forty-five parks have been successful.

Table 4-6 ranks the parks by their success scores in Table 4-5. The table distinguishes parks that employ more than one hundred employees from those employing fewer than one hundred to account for the possibility that very small

Table 4-5. *Research Park Success Indicators*

NAME OF PARK	LOCATION	YEAR ESTABLISHED	Research Park County			Control Counties			DIFF (%)
			BEFORE (%)	AFTER (%)	DIFF1 (%)	BEFORE (%)	AFTER (%)	DIFF2 (%)	
Ada Research Park	Ada, OK	1960	-1.92	1.33	3.25	0.25	3.51	3.27	-0.02
Ann Arbor Technology Park	Ann Arbor, MI	1983	-1.03	3.67	4.70	-2.97	3.23	6.19	-1.49
Arizona State University Research Park	Tempe, AZ	1984	3.40	6.22	2.82	2.76	5.75	2.99	-0.17
Carolina Research Park	Columbia, SC	1983	0.90	3.31	2.41	0.98	3.54	2.57	-0.16
Central Florida Research Park	Orlando, FL	1979	5.88	7.92	2.03	5.28	6.59	1.31	0.72
Charleston Research Park	Charleston, SC	1984	2.24	3.49	1.25	1.02	3.47	2.45	-1.20
Chicago Technology Park	Chicago, IL	1984	-1.41	1.86	3.26	4.58	6.19	1.61	1.65
Clemson Research Park	Clemson, SC	1984	-0.85	1.59	2.44	1.61	3.67	2.06	0.38
Connecticut Technology Park	Storrs, CT	1982	2.00	6.17	4.18	3.23	4.23	1.00	3.18
Cornell Research Park	Ithaca, NY	1951	-2.62	1.45	4.07	-2.39	-7.80	-5.41	9.48
Cummings Research Park	Huntsville, AL	1962	2.71	7.40	4.69	-0.83	3.46	4.29	0.40
Engineering Research Center	Fayetteville, AR	1980	3.91	3.43	-0.48	3.09	2.60	-0.50	0.02
Great Valley Corporate Center	Malvern, PA	1974	2.73	3.95	1.21	-0.20	0.92	1.13	0.08
Innovation Center and Research Park	Athens, OH	1978	1.06	0.67	-0.39	1.67	-0.66	-2.33	1.94
Interstate Business Park	Tampa, FL	1983	4.68	6.60	1.92	4.06	7.03	2.97	-1.05
Johns Hopkins University Research Park	Baltimore, MD	1984	1.81	4.08	2.27	1.44	4.35	2.91	-0.64
Langley Research & Development Park	Newport News, VA	1966	3.84	-6.00	-9.84	4.79	3.75	-1.04	-8.80
Maryland Science & Technology Center	Adelphi, MD	1982	2.88	6.10	3.23	2.52	4.31	1.79	1.44
Massachusetts Biotechnology Research Park	Worcester, MA	1984	-0.93	4.06	4.99	1.25	2.76	1.51	3.48
Miami Valley Research Park	Kettering, OH	1981	1.59	3.06	1.46	1.04	2.38	1.35	0.11
Morgantown Industrial & Research Park	Morgantown, WV	1973	3.69	4.44	0.75	2.20	2.71	0.51	0.24

Name of Park	Location	Year Established	Research Park County			Control Counties			
			Before (%)	After (%)	DIFF1 (%)	Before (%)	After (%)	DIFF2 (%)	DIFF (%)
Ohio State University Research Park	Columbus, OH	1984	0.02	5.02	5.00	-1.66	2.56	4.22	0.78
Oregon Graduate Center Science Park	Beaverton, OR	1982	7.94	5.03	-2.91	2.13	3.54	1.42	-4.33
Princeton Forrestal Center	Princeton, NJ	1975	2.64	1.82	-0.82	2.21	4.90	2.69	-3.51
Purdue Industrial Research Park	W. Lafayette, IN	1961	0.15	5.96	5.81	-0.06	4.66	4.72	1.09
Rensselaer Technology Park	Troy, NY	1982	1.17	3.83	2.66	1.68	2.72	1.04	1.62
Research Forest	The Woodlands, TX	1984	11.16	0.24	-10.90	3.04	0.73	-2.31	-8.59
Research Triangle Park	R.T.P., NC	1958	-0.29	4.37	4.66	1.98	2.19	0.21	4.45
Richland Industrial Park	Richland, WA	1962	0.98	1.73	0.75	0.73	4.85	4.12	-3.37
Roswell Test Facility	Roswell, NM	1983	2.81	-1.09	-3.89	0.40	0.46	0.05	-3.94
Shady Grove Life Sciences Center	Rockville, MD	1976	4.30	3.85	-0.44	1.40	1.55	0.15	-0.59
Stanford Research Park	Palo Alto, CA	1951	8.09	8.37	0.28	6.05	2.80	-3.25	3.53
Sunset Research Park	Corvallis, OR	1983	-0.03	3.06	3.09	-3.06	2.33	5.39	-2.30
Swearingen Research Park	Norman, OK	1950	-2.33	7.37	9.70	0.21	3.63	3.42	6.28
Synergy Research Park	Richardson, TX	1982	6.24	3.89	-2.35	7.12	0.54	-6.58	4.23
Tennessee Technology Corridor	Knoxville, TN	1982	3.59	2.35	-1.24	2.48	4.61	2.13	-3.37
University Center R&D Park	Tampa, FL	1982	5.81	6.60	0.79	6.41	4.64	-1.76	2.55
University City Science Center	Philadelphia, PA	1963	0.32	1.78	1.46	-0.12	3.87	3.98	-2.52
University of California – Irvine Park	Irvine, CA	1983	4.17	6.36	2.19	2.11	4.17	2.06	0.13
University of Utah Research Park	S.L.C., UT	1970	2.96	6.15	3.19	5.93	5.45	-0.48	3.67
University Park	Mt. Pleasant, MI	1982	3.38	3.62	0.25	0.61	3.47	2.86	-2.61
University Research Park	Charlotte, NC	1968	14.10	3.96	-10.10	8.97	2.41	-6.55	-3.55
University Research Park	Madison, WI	1984	0.54	3.92	3.38	-0.67	2.43	3.10	0.28
Washington St. Univ. Res. & Tech. Park	Pullman, WA	1981	0.34	2.07	1.73	3.81	1.94	-1.87	3.60
Westgate/Westpark	McLean, VA	1982	1.46	10.12	8.65	5.63	6.13	0.49	8.16

Table 4-6. *Ranking of Parks by Success Indicators* (N = 45)

Parks Successful Using MSR 120	%
Cornell Research Park	9.48
Westgate/Westpark	8.16
Swearingen Research Park	6.27
Research Triangle Park	4.45
Synergy Research Park	4.23
University of Utah Research Park	3.66
Washington State University Research and Technology Park	3.60*
Stanford Research Park	3.53
Massachusetts Biotechnology Research Park	3.48
Connecticut Technology Park	3.17
University Center R&D Park	2.55*
Innovation Center and Research Park	1.94*
Chicago Technology Park	1.66
Rensselaer Technology Park	1.62
Maryland Science and Technology Center	1.44
Purdue Industrial Research Park	1.09
Parks Successful Using MSR 100	
Ohio State University Research Park	0.77
Central Florida Research Park	0.72
Cummings Research Park	0.40
Clemson Research Park	0.39*
University Research Park (WI)	0.28
Morgantown Industrial and Research Park	0.23
University of California – Irvine Park	0.12*
Miami Valley Research Park	0.12
Great Valley Corporate Center	0.09
Engineering Research Center	0.02*
Unsuccessful Parks	
Ada Research Park	−0.02
Carolina Research Park	−0.15
Arizona State University Research Park	−0.17
Shady Grove Life Sciences Center	−0.59
Johns Hopkins University Research Park	−0.65
Interstate Business Park	−1.05
Charleston Research Park	−1.20
Ann Arbor Technology Park	−1.49
Sunset Research Park	−2.30
University City Science Center	−2.53
University Park (MI)	−2.61
Richland Industrial Park	−3.37
Tennessee Technology Corridor	−3.37
Princeton Forrestal Center	−3.51
University Research Park (NC)	−3.54
Roswell Test Facility	−3.94
Oregon Graduate Center Science Park	−4.33
Research Forest	−8.60
Langley Research and Development Park	−8.79

* Park has fewer than 100 employees.

parks are not likely to have had a significant impact on their region, despite the high success index they may have been assigned.

The table lists sixteen parks that are judged to have been successful under our stringent criterion, another ten parks that have been successful under our more lenient criterion, and nineteen parks that have been unsuccessful—at least in terms of induced total employment growth. Six of the parks shown to be successful have fewer than one hundred employees. It is possible that these parks simply are too small to have been engines of growth in their regions. We also could have calculated an erroneously large success indicator by not controlling completely for county size in our quasi-experimental research design. However, two of these parks are in nonmetropolitan areas and three are in metropolitan areas with fewer than 500,000 people. In at least five of the six cases, then, it is plausible that small parks may have stimulated positive employment growth.[11]

What accounts for these rankings? Specifically, are there local economic and/or park characteristics that systematically explain a park's success or lack of success by our measures? First, it is critical to stress that the definition of success as used here is a limited one. One limitation of our approach is that we looked at the employment growth rate for only the first five years after park creation. Many of the economic development impacts of parks will take longer than five years to materialize. In addition, parks that are not ranked high in our lists may still be judged to be successful by other measures. Conversely, because we were unable to control for *all* conceivable rival factors, we undoubtedly have listed as successful some parks whose counties would have been growing relative to their respective control groups even without a research park.

Table 4-7 presents cross-tabulations of the success indicators and four key characteristics of research parks: (1) the parks' vintage, (2) the geographic region of the county in which the park is located, (3) the size of the population of the metropolitan area in which the park is located, and (4) the type of university with which the park is formally or informally associated. Vintage is important because it takes time for a park to establish linkages with other businesses in the region, because new R&D organizations are highly attracted to regions that already have a concentration of R&D (that is, localization economies matter to R&D organizations), and because there is a premium for being an early bird since the supply of technologically oriented businesses is limited. Geographic region was included as a proxy for the local industrial base and political culture. For example, the northeastern and north-central/midwestern regions

Table 4-7. *Park Success and Selected Characteristics*

	Characteristics	Number of Parks	Year Established[*]	Average Difference	% Successful MSR 100	% Successful MSR 120
Vintage	Old	12	1950–70	0.009	58.3	50.0
	Middle–aged	18	1973–82	0.007	72.2	55.6
	Young	15	1983–84	-0.009	40.0	13.3
Region	Northeast	7	1975.0	0.017	71.4	57.1
	North–Central/Midwest	8	1982.5	0.002	75.0	37.5
	South	22	1982.0	0	50.0	36.4
	West	8	1981.5	-0.004	50.0	37.5
Size (1980 population range)	>2.5 million	7	1982.0	-0.001	57.1	42.9
	1 mil. – 2.5 mil.	11	1982.0	-0.001	54.5	36.4
	500,000 – 999,999	9	1979.0	0.015	77.8	66.7
	100,000 – 499,999	10	1981.5	-0.008	50.0	10.0
	<100,000	8	1979.5	0.008	50.0	50.0
Type of Research University[**]	Type I	15		0.006	60.0	40.0
	Type II	8		0.004	62.5	50.0
	No affiliation	22		0	54.5	36.4

[*] The entries for "Region" and "Size" are median years.
[**] The two types of research universities are defined in n. 12 below.

generally have older manufacturing bases and higher rates of unionization than the southern and western regions. Size was included to capture the presence of agglomeration and urbanization economies. Finally, the type of university enabled us to test whether the ability for neighboring private sector scientists/engineers and university researchers to collaborate had affected growth.

Our analysis of the data displayed in Table 4-7 suggests that the four key characteristics of research parks play an influential role in the extent to which they contribute to regional economic development:

Vintage. We split the parks into three vintages. Because no parks were established between 1971 and 1973, we made the break between old and middle-aged parks at that point. Thus, old parks have existed for at least twenty years; middle-aged parks, for at least eight years; and young parks, for no more than eight years. Old and middle-aged parks indeed appear to have been more successful than the youngest group of parks. The difference in performance between old and middle-aged parks is not large and may be an artifact of the arbitrary dividing line we have drawn.

Region. We can make two observations of note from the left-hand side of the table: parks in the Northeast generally are older than those in other regions, and almost half of all parks are in the South. Entries on the right-hand side of the table suggest that parks in the Northeast and the North-Central region have been more successful than those in the South and the West.

Metropolitan Area Population. Parks in medium-sized regions, with populations between 500,000 and 1,000,000, appear to have performed better than other parks, and parks in small areas, with populations less than 100,000, have performed better than many might have expected. These results, of course, are sensitive to how we arbitrarily drew up the size classes.

The results for small areas can be explained, in part, by the fact that parks located in these regions can serve the same function as a central business district: they can be a source of agglomeration economies that small places otherwise would lack. That parks in areas with populations between 500,000 and 1,000,000 have performed relatively better might be explained by the fact that those areas are sufficiently large to offer the various urbanization and agglomeration economies that attract R&D organizations. Such economies include a diversified pool of skilled labor, cultural amenities, good airline service, and necessary business support services. Yet those areas are not so large as to have generated congestion, environmental degradation, a high cost of living, and other diseconomies of metropolitan scale.

Affiliation with Research Universities. Parks affiliated with Type I research universities appear to have been more successful than parks without that affiliation.[12] There is no clear difference between parks associated with Type II research universities and parks without a research university affiliation. This may be because the counties shown to have no affiliated research university may still have doctoral-granting universities, specialized engineering and medical institutions, and other types of higher-education facilities that also can benefit businesses in research parks.

One of the difficulties in interpreting the results in Table 4-7 is due to the fact that the effect of each causal factor on success is not isolated from all other factors. A standard way to control for other factors is to employ multiple regression analysis. We have performed this type of analysis in which we explain the variation in the park success measure (that is, the dependent variable) by the characteristics listed in Table 4-7, as well as other explanatory variables.[13] Alternative measures for seven types of factors were formulated and tested as well, including (1) location, (2) vintage, (3) characteristics of park businesses, (4) university linkages, (5) park-provided services, (6) park-imposed restrictions, and (7) governmental assistance.

Three alternative measures of *locational characteristics* were developed: a regional dummy (or indicator variable) to capture industrial base, political, and sociocultural differences; a dummy to indicate whether the park's county was a core county in a metropolitan area, a metropolitan noncore county, or a nonmetropolitan county; and the park region's population. The last two measures are highly correlated and serve as proxies for the same underlying phenomena—namely, the presence of agglomeration and urbanization economies. Consequently, they (and others that are similarly correlated) were not used in the same regression model.

For *vintage* we used the number of years that parks had existed before 1985. Because 1951 was our first observation, this variable ranged from 1 to 34. As *characteristics of park businesses*, we used data from Table 4-3 on the types of favored businesses within parks and the percentage of floor space within parks that is in incubators for small businesses. We tried three measures of *university-*

Table 4-8. *Ordinary Least Squares Regression Results* (N = 45)

VARIABLE	COEFFICIENT	STANDARD ERROR	SIGNIFICANCE LEVEL
Constant	−0.0309	(0.017)	0.067
Northeast (dummy)	0.0135	(0.019)	0.490
North–Central/Midwest (dummy)	0.0171	(0.018)	0.358
South (dummy)	0.0128	(0.016)	0.422
Vintage squared	2.87 E-5	(1.43 E-5)	0.050
MSA population	2.43 E-9	(3.95 E-9)	0.549
Deed restrictions used (dummy)	0.00488	(0.014)	0.730
Garbage collection provided (dummy)	0.0221	(0.013)	0.096
Government subsidies provided (dummy)	−0.0115	(0.012)	0.350
University–owned (dummy)	0.0227	(0.012)	0.072

Dependent variable = 0/1, based on value of DIFF in Table 4-5.
R-squared = 0.28; Adjusted R-squared = 0.10.
F-statistic (9,35) = 1.5008; Significance of F–Test = 0.185.
Significance level = 0.146.

park linkage: a dummy indicating whether the parks were owned and operated by a university, a dummy indicating whether the parks were proximate to a research university, and a dummy indicating whether the parks were near a Type I research university. The *park-provided services* we included as dummies were garbage collection, fire protection, and road maintenance. The *restrictions* we tested, which are from Table 4-2, included deed restrictions generally and a limitation on manufacturing activity. Finally, we included dummy variables (0 or 1) in different regressions to indicate whether park businesses or management had received *government subsidies*.

Three of the variables listed above proved to be statistically significant explanatory factors in most of the alternative models that were estimated: the age, or vintage, of the park; formal affiliation with a public or private university; and the provision of garbage collection services. In other words, these are the factors best able to explain the variations in the measures of research park success. Presented below are the results from two of the models (ordinary least squares and logit models), followed by additional results from a hazards/survival model.

Table 4-8 contains the results from an ordinary least squares regression of the relative employment growth rate differences ("DIFF" in Table 4-5) on re-

Table 4-9. *Logit Model Results* (N = 45)

VARIABLE	COEFFICIENT	STANDARD ERROR	SIGNIFICANCE LEVEL
Northeast (dummy)	–0.138	(1.520)	0.927
North–Central/Midwest (dummy)	2.477	(1.537)	0.107
South (dummy)	0.581	(0.001)	0.615
Vintage squared	3.75 E–4	(1.20 E–3)	0.755
MSA population	3.55 E–7	(3.22 E–7)	0.270
Deed restrictions used (dummy)	–1.625	(1.137)	0.153
Garbage collection provided (dummy)	3.000	(1.266)	0.018
Government subsidies provided (dummy)	–1.736	(1.067)	0.104
University–owned (dummy)	3.068	(1.352)	0.023

Dependent variable = 0/1, based on value of DIFF in Table 4–5.
Log–likelihood statistic = –19.813.
Restricted (slopes = 0) log–likelihood statistic = –30.645.
Chi–squared (8) = 21.664.
Percent successes predicted = 76.1.
Significance level = 0.006.

gion (represented by dummy variables), the square of vintage, metropolitan population, the use of deed restrictions (represented by a dummy variable), the provision by park management of garbage collection services (dummy variable), the use of government assistance by park businesses (dummy variable), and park ownership by a private or public university (dummy variable). We squared the vintage variable for two reasons. First, the results in Table 4-6 suggest that vintage (time) enters the model nonlinearly. Second, we wanted to count time more heavily than other variables in the analysis because, as discussed above, it can contribute to success in several ways. Garbage collection was used as the bellwether infrastructure service because it is one of the few services for which there are private alternatives; hence, provision of garbage collection services by park management represents a convenience and probable cost savings to park businesses. Finally, after trial and error, the organizational status of the park (owned by a university) was chosen as the measure of "the university connection." We believe that it outperforms other measures of university affiliation in our models because it is less ambiguous.

The results indicate that the vintage, garbage collection, and university variables each are significant explanatory variables (at the 0.10 level of significance)

Table 4-10. *Hazards/Survival Model Results* (N = 45)

Duration (Years)	Number Entering	Number Censured (N = 26)	Number at Risk	Number Exiting (N = 19) (%)	Hazard Rate (SE)[*]
0 – 3.8	45	0	45	0 (0)	0 (0)
3.8 – 7.6	45	14	38	12 (31.6)	0.099 (0.03)
7.6 – 11.4	19	3	17	0 (0)	0 (0)
11.4 – 15.2	16	2	15	2 (13.3)	0.038 (0.026)
15.2 – 19.0	12	1	11	0 (0)	0 (0)
19.0 – 22.8	11	0	11	2 (18.2)	0.053 (0.037)
22.8 – 26.6	9	1	8	2 (23.5)	0.07 (0.049)
26.6 – 30.4	6	2	5	1 (20.0)	0.059 (0.058)
30.4 – 34.2	3	0	3	0 (0)	0 (0)
34.2 – 38.0	3	3	1	0 (0)	0 (0)

[*] The standard error for the hazard function is: $SE(x_j) = x_j[(1-hx_j/2)^2]/(r_j q_j)^{1/2}$.

and relate to the dependent variable in the expected direction. The overall explanatory power of the model, however, is low. This is not surprising given the nature of the data and the small sample size.

Table 4-9 shows the results from a regression model using a dichotomous measure of park success as the dependent variable and logit estimation.[14] Here, garbage collection and university ownership are statistically significant variables, but the vintage variable ceases to be a significant explanatory factor of park success.

Because the results of the ordinary least squares and logit models were inconsistent for the vintage variable, we estimated a third, hazards/survival model (see Table 4-10). This class of models is used with events (such as the creation of research parks) that can terminate in death (or failure) over time. The model allows one to estimate survival rates (or, in this context, success rates, as indicated in Table 4-5) for different vintages. Table 4-10 shows ten time periods, the number of parks established during those periods, the number of parks in

Table 4-11. *Hazards/Survival Model with Covariates*

Variable (covariate)	Coefficient	Standard Error	Significance Level
Garbage collection provided (dummy)	–0.988	(0.75)	0.189
University–owned (dummy)	–2.077	(1.03)	0.044

Log–likelihood statistic = –54.924.
Restricted (slopes = 0) log–likelihood statistic = –59.597.
Chi–squared (2) = 9.347.
Significance level = 0.009.

each period that are "censured," the number of parks at risk and exiting, and the hazard rate. We equated failures (as described above) to "exits." Parks that we have identified as successful were censored—that is, marked as potential failures.[15] The key insight from the table is that hazard rates are higher and thus more significant for younger parks than for older parks.

The hazards/survival model can be extended to estimate how various other factors (that is, covariates) affect the hazard rate.[16] We introduced garbage collection and university ownership as covariates in this analysis because they were shown to be significant in the foregoing analysis. The results from this model are contained in Table 4-11. The negative signs on the coefficients for the two covariates included indicate that the provision of key services and university ownership and operation of parks reduce the hazard (that is, the likelihood of failure).

A Focus on Park Failures

In our discussion of the econometric results above, we focused on the determinants of success, measured as the direct plus induced regional employment growth that can be attributed to research parks. The results also shed some light on the determinants of failure, at least to the extent that we define failure as a small or null value for the dependent variable. For example, we can interpret the ordinary least squares regression results in Table 4-8 and the hazards model results in Table 4-10 to mean that the younger the park, the higher the degree of failure. Similarly, the results in Table 4-9 can be read to suggest that parks that do *not* collect garbage are more likely to fail than parks that do provide that service.

The insights these results give about park failures are limited for at least two reasons. First, the data used in the regressions are only from parks that have not ceased to operate. We also would like to know how the parks that have "died" differed from those that continue to live. Second, the particular measure of success we have used is most appropriate for parks that have passed into the maturation stage for the very fact that induced employment effects can take years to materialize. Consequently, younger parks in the sample that have not yet entered the maturation stage might be judged as failures prematurely. Ideally, we would identify predictors that indicate which young parks are likely to proceed into the maturation stage and which parks will not.

A complete empirical analysis of failure is not possible because data are difficult—and, in many cases, impossible—to collect from parks that have ceased to exist. However, we can use anecdotal information about failed parks and data from existing parks to understand better how parks that fail differ from parks that are successful. In the case of a park operated by a large research university in a nonmetropolitan part of a western state, many of the more tangible requisites of a successful research park were in place, including available land with good amenities, a university with research strengths in engineering and the sciences, and up-front funds for infrastructure. Nonetheless, the park only attracted a few small tenant organizations; after a number of years of operation, there were fewer than one hundred employees in the park. The park director cited the lack of effective leadership and commitment by top university officials as a reason for the park's poor performance. University officials just assumed that physically locating a research park in close proximity to university faculty and other researchers almost automatically would lead to close university-industry collaborations that, in turn, would attract additional private R&D organizations to the park. The park director believed that university leaders did not fully appreciate the different subcultures of industries and universities and did not provide sufficient incentives and institutional support to induce faculty to work with private industry. Such incentives and support might include weighting collaborative work by faculty more heavily in the tenure and promotion process, providing salary supplements to encourage that kind of activity, or adding new faculty positions to departments willing to collaborate with industry.

Park managers and other research park professionals with whom we spoke, including some individuals who have worked in parks that did not attract businesses, identified two factors that are associated with park failure: (1) a lack of commitment and patience by key individuals, and (2) distributional politics. In one case, the state legislature withdrew funding support because "not much had happened" in the two years the park had been in existence. That attitude was likened to the child who kept digging up the ground where he had planted the flower to find out why the flower was not yet poking through the ground.

Representatives from two research parks associated with different branches of the same university in a southern state illustrated the problems created by distributional politics. The flagship campus already had a research park. Other branches of the state university lobbied successfully for authorization to build their own parks. That not only diluted the state funds that might have been available to support the research park at the flagship campus, but it also reduced each park's chances of landing one of the relatively few R&D organizations that indicated some interest in locating in the state. As a consequence, the ability of each park to grow to a threshold size with a critical mass of scientists, engineers, and entrepreneurs was diminished.

Data from existing parks reveal another important set of reasons for failure: (1) inadequate population size and growth potential in the region, and (2) the absence of a research university. If a region is small, its size limits the possibilities of developing agglomeration and urbanization economies. Slower growth means that the region has relatively less future development potential, including expanding pools of labor and supplies. Not having proximity to a research university restricts R&D organizations' access to intellectual capital, regardless of the region's size.

Table 4-12 displays data from three types of parks: those with no employees, those with less than one hundred employees, and those with one hundred or more employees. The data include the average metropolitan area size and growth rate, the percentage of parks in metropolitan areas, and the percentage of parks affiliated with research universities, by year of research park birth, for each employee size class. Based on the analysis of success and failure presented in Chapter 3, we can say that parks established before 1985 that had zero employment as of 1988 were failures in the incubation stage. Similarly, parks established before 1982 that had fewer than one hundred employees as of 1988 can be designated as failures in the consolidation stage, and parks established before 1985 that had fewer than one hundred employees as of 1988 can be classified as "at risk of failure" in that second stage of development.

We can use the table to ask: How do the parks that have failed differ from

Table 4-12. *Park Characteristics, by Size and Vintage*

	0 Employees (N = 17)		
	Failure in Stage 1		*At Risk in Stage 1*
	ESTABLISHED BEFORE 1982 (N = 5)	ESTABLISHED 1982–84 (N = 4)	ESTABLISHED 1985–88 (N = 8)
Average MSA population	520,666	717,438	583,345
% in metro areas	80%	75%	88%
Average area growth rate	2.74%	4.80%	4.83%
% with type I, II research universities	40%	25%	50%
	1–99 Employees (N = 28)		
	Failure in Stage 2	*At Risk in Stage 2*	*(Parks Not Yet in Stage 2)*
	ESTABLISHED BEFORE 1982 (N = 4)	ESTABLISHED 1982–84 (N = 10)	ESTABLISHED 1985–88 (N = 14)
Average MSA population	421,191	1,216,956	1,574,793
% in metro areas	75%	80%	71%
Average area growth rate	4.69%	5.18%	4.08%
% with type I, II research universities	50%	50%	43%
	100 Employees or More (N = 75)		
	Successful in Stages 1 or 2; Perhaps in Stage 3		
	ESTABLISHED BEFORE 1982 (N = 29)	ESTABLISHED 1982–84 (N = 26)	ESTABLISHED 1985–88 (N = 20)
Average MSA population	1,079,945	1,616,724	990,540
% in metro areas	83%	88%	90%
Average area growth rate	5.49%	5.55%	5.7%
% with type I, II research universities	62%	71%	70%

those in the table that have not? Parks that have failed are more likely to be located in smaller regions and in counties that have had lower employment growth rates, and they are less likely to be associated with a research university. Each of these observations is consistent with regional development theory as presented in Chapter 2.

Summing Up the Evidence

In this chapter we have presented several pieces of evidence about the success and failure of research parks, measured in terms of their ability to generate jobs directly and induce employment growth. We noted, first, that as many as one-half of all parks that are announced fail in the incubation stage or early in the consolidation stage. Many of those that survive are converted from research to more general business parks. The latter group of parks may be successful as real estate ventures but not as policy undertakings, since their original objective (to attract a critical mass of R&D activity) has not been achieved. We showed, finally, that of the extant parks, only about half can be judged successful in terms of their ability to create jobs that otherwise probably would not have existed in the region.

Next, we noted that to the extent that park managers believe that parks have affected *any* economic development outcomes, those outcomes have been growth related. The quasi-experimental and econometric analyses presented indicate some of the critical factors that are necessary for parks to stimulate significant employment growth directly and indirectly within their regions. These factors can be summarized as vintage, linkage, and amenities. Vintage refers to the date the park was established. Our results suggest that "the early bird gets the worm," since the probability of success is higher for earlier parks than for later ones. The results also indicate that the nature of the university-park linkage matters, since regions with parks that also contained research universities had better employment growth than regions that did not. Finally, our results suggest that amenities matter. All else being equal, businesses favor research parks that provide essential services. The greater the provision of these services, the more attractive parks are to businesses and the greater the job creation both within and outside the park.

These results are substantiated when park failures are compared to successes controlling for vintage. Failing parks are more likely than successful parks to be located in smaller regions and in slower growing counties and not to be related with a nearby research university. Anecdotal evidence about parks that have failed stresses the lack of patience, commitment, and understanding of appropriate university-industry institutional relationships by key leaders and the propensity for legislatures and university governing boards to approve too many parks.

Our findings have implications for both theory and policy. The implications for policy are discussed in Chapter 9. The conclusion we draw for theory is that research parks do not behave as a classic growth/development pole in larger regions that contain research universities, since backward and forward linkages do not play a major role in the generation of employment in those areas. The universities themselves seem to be acting as the growth poles, with localization economies, rather than linkages based on material inputs and outputs, providing most of the growth stimulus.

In regions without research universities and in smaller metropolitan areas that otherwise lack the basis for agglomeration economies, research parks can serve as a growth/development pole. But the strength of the pole in those instances depends on the other critical factors mentioned above—vintage and amenities—as well as on intangibles, including good fortune and wise and effective leadership.

5 The Research Triangle Park

Created in 1959, the Research Triangle Park (RTP) in North Carolina is the largest, and is considered to be one of the most successful, research parks in the world. RTP occupies 6,700 acres in the middle of a triangle formed by the University of North Carolina in Chapel Hill, Duke University in Durham, and North Carolina State University in Raleigh. There are approximately fifty R&D-oriented organizations in RTP with a combined workforce of about 32,000. From its beginning, the park has deliberately sought the R&D branch plants of major, technology-oriented corporations. The list forms a veritable who's who of the Fortune 500—Data General, Dupont, IBM, Northrop—as well as foreign-based firms such as BASF, Burroughs-Wellcome, Ciba-Geigy, Glaxo, Northern Telecom, Rhone-Poulenc, and Sumitomo. These and other organizations occupy spacious, low-rise, often architecturally distinct buildings in a low-density, wooded setting. Indeed, the appellation *park* is no misnomer in the case of RTP.

Each year, hundreds of economic development officials from all over the United States and many parts of the world visit the Research Triangle Park to understand how the park works and to see if it can be adapted to their own regions. In addition to being one of the earliest, largest, and most parklike planned concentrations of R&D activity in the world, RTP serves as a model because it is the symbol of one of the most dramatic cases of regional economic restructuring that has yet been documented. For that reason, representatives of regions that have little or no technology-oriented activity or tradition look to RTP and see some reason for optimism. The story of RTP is also one in which a particular set of actors has made an important difference.

Historical Background[1]

In the mid-1950s, North Carolina had the second lowest per capita income of any state. Its employment base was concentrated in three low-wage, declining industries: tobacco, textiles, and furniture. There was little or no R&D activity in the state except for that in the three research universities (University of North Carolina at Chapel Hill, Duke University, and North Carolina State University). The combination of high-quality research universities and the lack of job opportunities for highly skilled scientists and engineers had led to a brain drain from the state of serious proportions.[2]

In 1955, Governor Luther H. Hodges formed a committee of state business leaders and prominent university officials to investigate how the strengths of the research universities could be used to help restructure North Carolina's economy. One year later the committee produced a report stating that the three research universities could attract a concentration of industrial research laboratories to the region to take advantage of faculty expertise in particular fields of strength. Economic development would spread to surrounding parts of the state by production facilities wanting to locate in some general proximity to their R&D labs. It is noteworthy that this report did not envision a research park. Because only a few, relatively new research parks existed at that time, the concept was not yet well known to many North Carolina officials.

Not much happened until two years after the report was issued, when Karl Robbins, a retired industrialist, was recruited to invest in the concept introduced by Governor Hodges's committee.[3] He proposed to build a private research park on 4,000 wholly undeveloped acres near the small airport in the center of the triangle formed by the three cities. Robbins failed to attract sufficient additional investors, in part because of public skepticism about the research park concept and in part because questions of propriety were being raised about the promotion of a privately owned research park by public universities and other state government agencies.[4] After another period of stagnation, a group of private citizens and civic-minded corporations in the state bought out the stock of the (empty) private research park and formed the Research Triangle Committee, Inc., in 1956. It hired George L. Simpson, Jr., a protégé of Howard Odum and a senior staff member of the Institute for Research in Social Sciences of the University of North Carolina, to develop a plan and to promote the area as a location for industrial R&D labs. After raising funds, the committee changed its name to the Research Triangle Foundation to govern the renamed

Research Triangle Park. A $500,000 grant and a gift of 157 acres in the middle of the park by the foundation were used to create a nonprofit contract research organization—the Research Triangle Institute—as the park's first occupant.

The park was slow to attract additional organizations until 1965, when IBM bought a large site for a major facility. Very shortly after IBM made its decision, Luther Hodges, who had gone from the governorship to Washington, D.C., as President John F. Kennedy's secretary of commerce, and Terry Sanford, who was then governor, were instrumental in getting the National Institutes of Health to locate its National Institute of Environmental Health Sciences (NIEHS) in the park as well.[5] These two occupants served as anchors by putting RTP on the map as a place to locate R&D facilities. A string of R&D branch plants of large national and multinational corporations and of federal government laboratories followed over the next twenty-two years. Since 1987, the growth of new enterprises locating in RTP has slowed as a result of a falloff in the rate of expansion of new corporate R&D facilities nationally, increased competition from other regions bidding for R&D activity (for example, MCC and Sematech locating in Austin, Texas), and the Research Triangle Foundation's delay in investing in the infrastructure needed to open up the undeveloped portion of the park.[6] Also, throughout the past three decades RTP management's promotional strategy has been to place the highest priority on attracting the branch plants of major corporations rather than new, small, start-up, technology-oriented businesses. Because of zoning and stringent building and site restrictions and the only very recent investment in multitenant "spec" buildings (except for offices and service functions), it has been economically infeasible for most small, start-up businesses to locate in the park. This may be another reason why the growth of RTP—in terms of the number of enterprises—has slowed.

Local Conditions and Resources

The Research Triangle region consists of three counties—Durham, Orange, and Wake—that together had a population of 324,000 and total employment of 124,300 in 1960.[7] The region has experienced a high growth rate in the last thirty years, particularly since around 1970. By 1986 the area had reached a population of 614,800 and total employment of 286,900.[8] But in the late 1950s, the area was a small metropolitan region with few of the agglomeration and urbanization economies that characterized larger metropolitan regions.

Thirty years ago, the Raleigh-Durham area was not only small but it also was peripheral. The area has been well-connected by the interstate highway system on the north-south axis for many years, but it is still located about 250 miles from Washington, D.C., and about 380 miles from Atlanta, the Southeast's regional center. The Raleigh-Durham airport now serves as one major airline's regional hub and has over 275 flights per day, including non-stop flights to Europe, the Caribbean, and Mexico, but in the late 1950s and 1960s, air connections to major business centers in the United States were still rather poor.[9]

The area's employment base in 1959 was concentrated in low-wage manufacturing industries (textiles, tobacco, and furniture), marginal farming, state government, and education. The latter two sectors helped make the area better off than the state as a whole in terms of its wage and salary levels, general working conditions, and economic stability. Yet there was still a marked absence of high-tech employment sectors (3.3 percent in the Raleigh-Durham area compared to 10.3 percent for the United States) and little or no tradition of entrepreneurial activity.[10] The resident labor force was disproportionately concentrated in professional and administrative occupations, employed largely in the universities and state government, on the one hand, and in semiskilled and low-skilled occupations, on the other. It was underrepresented in the skilled trades and technical occupations. While the percentage of area residents that had attended college was well above the national average (24.2 percent versus 16.5 percent), the median number of years of school completed was only slightly above the national average (10.8 versus 10.6 years). The proportion that had not finished high school was 56.6 percent in 1960, compared to 67.7 percent for the state but 58.9 percent for the nation.[11] An ambitious investment in a large number of community colleges had begun under Governor Luther Hodges in recognition that the lack of technical training was a barrier to economic development in the state. Despite this investment, the problem of a scarce number of skilled technicians and, more generally, a poorly educated resident labor force persisted through the 1980s.[12]

There is little question that one of the two principal resources of the area lay within its three research universities. In particular, the University of North Carolina's chemistry department, with a national reputation in organic and biochemistry, had a long tradition of supplying chemists with graduate degrees to industry, including the laboratories of the nation's and world's major chemical corporations. That, combined with North Carolina State University's highly regarded School of Textiles, explains RTP's success in attracting, and developing an early concentration in, textile chemistry R&D labs. Later, the strengths of the biomedical research faculty and facilities of the University of North Carolina and Duke University and the strengths of North Carolina State University's agricultural sciences faculty became instrumental in attracting pharmaceutical and biotechnology research labs to the park. Likewise, the engineering schools at North Carolina State and Duke and the computer science department at the University of North Carolina paved the way for microelectronics R&D facilities to locate in the park.

Today, the three universities combined annually graduate approximately 2,500 students with master's degrees and 725 with Ph.D. degrees in arts and sciences, engineering, textiles, the agricultural sciences, forestry, and business.[13] Hundreds more receive advanced degrees from North Carolina Central University in Durham, the University of North Carolina at Greensboro, Wake Forest University in Winston-Salem, and East Carolina University in Greenville. The last four institutions are less than one hundred miles from the park. From the point of view of a prospective industrial R&D lab, direct access to these graduates for recruiting purposes represented a significant comparative advantage for the area compared to the rest of the South.

The second principal resource of the area was a line of governors, who by southern standards were politically moderate and activist (including those whom Luebke refers to as "modernizers"), and a group of civic-minded business leaders.[14] North Carolina's political leaders understood both the risks of waiting before trying to reverse the direction in which the state economy was headed and the key role that investments in higher education had to play in restructuring the economy; moreover, they realized that the fruits of their political investments would not be seen until well after their terms had expired. A group of business leaders that included many prominent industrialists from the state's traditional sectors as well as bankers also understood that economic diversification and modernization, a course that might not be in the best short- or medium-term interests of their individual companies, was necessary for the state's economic future. Later, when the Research Triangle Park (private at the time) was threatened for lack of investors, this group's ability to gain consensus on a course of action and to muster sufficient resources to implement it proved to be decisive.[15]

In summary, local conditions at the time seemed to tip against the probable success of RTP as an instrument for regional economic restructuring. On the positive side were the combined resources of the region's three research universities, which distinguished the area from any other in the South, and the vision and leadership of the state's political, business, and civic elite. On the other hand, the area had few of the resources commanded by large metropolitan agglomerations, including an abundant supply of local capital available for new business development. It occupied a peripheral location, did not have a strong manufacturing production base outside of the nondurable, low-wage traditional sectors, had little entrepreneurial traditions, and, except for university graduates, had a relatively low-skilled and poorly educated resident workforce. If one were to imagine informed experts placing wagers in the late 1950s on which areas of the United States would emerge as high-technology centers thirty years later, Raleigh-Durham probably would not have been on the list.

The Operations and Policies of the Park

RTP is owned and managed by the nonprofit Research Triangle Foundation. The foundation's president is the chief executive officer (CEO) and reports to a thirty-member governing board. Members of the board, which is self-perpetuating, include the governor of the state (or his/her representative), the president of Duke University, the president of the University of North Carolina system, and the president of the Owners and Tenants Association of the Research Triangle Park.

The foundation sells lots to organizations wishing to locate in the park. A small number of lots have been sold to developers who, in turn, lease building space. Land inside the developed part of the park currently sells for about $45,000 per acre. Pricing policies are set to meet a desired rate of return, rather than to match the current market price for comparable land outside the park. It turns out that the price of land in the park is somewhat below that of prime land sites outside the park on a per-acre basis, but minimum lot size restrictions (eight acres) and a maximum footprint of 15 percent make the cost of owning a building lot in RTP considerably more expensive than it is outside the park.[16]

Provision of most services is the responsibility of the organizations that locate in the park. The park management shares the burden of building and maintaining park roads, supplies signage, provides land use planning for the common and undeveloped areas, and acts as a liaison with local and state government agencies for the provision of public services to occupants. State government builds and maintains access roads to the park, shares the cost of roads in the park, and occasionally contributes Department of Commerce personnel and resources to efforts to recruit businesses to the park. Local government provides water and sewer services and police protection. Fire protection is supplied from volunteer fire departments. There is a hotel, several restaurants, and other selected services in the park, but these operate as normal "pay-for-services" businesses. Most of the organizations in the park supply their own eating facilities, private security service, means of waste and garbage disposal, landscaping and grounds maintenance, meeting facilities, and communications equipment.

The park finances its own operations with revenue from the sale of land. There are no direct government subsidies to RTP. The state and local governments have assisted the park by providing the traditional services mentioned above. These services are not available at a discount, however.[17] Perhaps the most important state government action was the creation, in 1984, of a special tax district for the park. That legislative act prevents RTP from being annexed, and property taxes from being assessed, by any nearby city. As a result, park landowners pay property taxes only to the county (mostly to Durham County) and not to both county and city, as owners of land in annexed areas must do. Indirectly, the federal government has assisted RTP by locating federal R&D labs there.

At the time that the Research Triangle Foundation was formed, the most important objectives for the park were to diversify the state's economic base, to increase the number of high-paying jobs in the state and substate region, and to increase employment opportunities for local university graduates and for North Carolinians who had left the state to go to college. Considered least important were the objectives to encourage entrepreneurship in the region and to increase technology transfer to existing businesses. As a nonprofit entity, the foundation places relatively low importance on maximizing revenue from the development and sale of land. Over the years, the ability of RTP to enhance the universities' training and research capacities and to increase employment opportunities for local university graduates has gained in relative importance. Perhaps because park officials perceive that regional economic diversification has been accomplished, the latter also has become less important as an objective of the park.

The foundation permits research and development, prototype manufacturing, and light manufacturing uses in the park (in addition to service functions). There is no limit on the amount of manufacturing conducted so long as it (1) is related to the R&D activity that is performed in the park facility and (2) takes place in areas that are zoned for "research applications," rather than just for research. For example, IBM now assembles in RTP *all* PS/2 microcomputers produced in the United States. The types of organizations that are most often targeted by park recruitment efforts are the R&D facilities of technology-oriented companies. There also is some preference for the branch plants of multilocational corporations and foreign-owned businesses. Start-up businesses are less favored. Park management says it has some preference for locally owned businesses, although thus far very few such businesses have located in the park.[18] RTP has no incubator facilities. And although there are now two multitenant spec buildings, the price per square foot of space and the lack of readily available venture capital in the region preclude the great majority of start-up businesses from locating in RTP.

Characteristics of R&D Organizations in the Park[19]

As mentioned earlier, RTP has specialized in the recruitment of R&D branch plants of large corporations. More than 70 percent of the organizations in the park are branches of multifacility organizations. Of those, only 37.9 percent have their headquarters in North Carolina, a low figure compared to both the Stanford and Utah research parks (71.4 percent and 61.5 percent, respectively).

Organizations in RTP are more oriented to basic and applied research than are firms in the University of Utah Research Park, which tend to emphasize product development or routine production more, but they are somewhat less oriented to basic or applied research functions than are organizations at the Stanford Research Park (see Figure 5-1). RTP organizations also have a somewhat larger proportion of scientists or engineers than Utah or Stanford organizations and a smaller proportion of skilled technicians (see Figure 5-2). Both the functional orientation and the occupational distribution of RTP organizations reflect their predominant status as R&D branch plants, with headquarters and production functions located at other facilities.

Proximity to the three research universities, access to highly skilled labor, and the quality of air service are the three most important reasons that organizations cite for deciding to locate in the region (see Table 5-1). These are consistent with the responses from the Stanford and Utah organizations as well. The next most important set of locational factors—the business climate, cul-

Figure 5-1. *Functional Specialization in the Research Triangle Park*

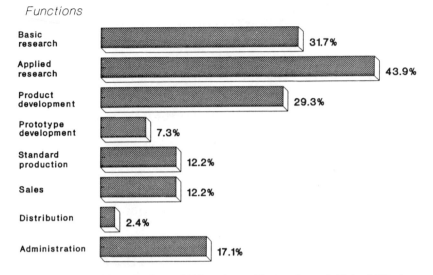

Functions

Percentage of Organizations with Function as First or Second Highest Effort
(N = 40)

Figure 5-2. *Occupational Mix in the Research Triangle Park*

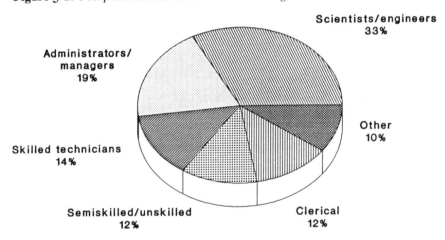

Percentage of Employment in R&D Organizations
(N = 40)

Table 5-1. *Reasons for Locating in the Research Triangle Region* (R&D Organizations in RTP, N = 40)

REASON	RANK IN IMPORTANCE
Proximity to research universities	1
Access to skilled labor	2
Access to major airport	3
Business climate	4
Cultural amenities	5
Physical climate	6
Quality of public services	7
Access to unskilled and semi–skilled labor	7
Access to markets	9
Access to business services	10
Quality and adequacy of infrastructure	11
Preference of CEO	12
Concentration of firms in same or related industry	12
Other branches in the region	14
Access to materials	15

tural amenities, and physical climate—is much more important at the Research Triangle Park, however, than at the other case study parks. On the other hand, factors associated with agglomeration economies, including the opportunity to develop linkages with other businesses in the area and to interact with other entrepreneurs and scientists, are relatively less important at RTP.

The prestige of being in RTP is the most important reason for choosing to locate in the park, given a location within the region (see Figure 5-3). Cited repeatedly in face-to-face interviews with CEOs, the prestige factor refers to the blue-chip quality of the large majority of firms located there. Opportunities to interact with professionals from other park organizations, or to share facilities, information, or other common inputs, are *not* important reasons for locating in RTP. Indeed, in the degree to which opportunities to interact were a factor in deciding to locate in the park, versus in a site outside the park, as well as in the actual frequency of interaction among RTP professionals in different

Figure 5-3. *Reasons for Locating in the Research Triangle Park*

Percentage of Organizations Citing as "Highly Important"

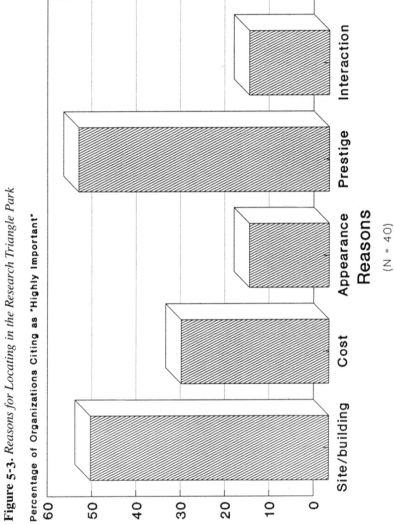

Reasons

(N = 40)

Table 5-2. *Importance of Research Universities to RTP Organizations*
(N = 40)

FACTOR	HIGH OR MODERATE IMPORTANCE (% of Respondents)
Access to entry–level labor force	85.7
Access to faculty consulting	71.4
Courses and training for employees	71.4
Cultural milieu	68.6
Opportunities to subcontract	67.7
Use of university facilities	55.9
Faculty appointments for staff	36.4

organizations, the difference between the Research Triangle Park and the other two case study parks is striking.

R&D organizations in RTP view their access to entry-level graduates for recruitment purposes as the most valuable benefit of the area's three research universities (see Table 5-2). The universities' contribution to a sociocultural milieu that is attractive to highly paid professionals, especially for those from outside the South, is a considerably more important factor here than in the cases of Stanford and Utah.

Impacts of the Park on Regional Economic Development

In this study, we have attempted to estimate the putative impacts of RTP on regional economic development, including firm location, employment growth, growth in the level of per capita personal income, level of income inequality, local labor market conditions of women and minorities, and overall innovative capacity of the region.

IMPACT ON FIRM LOCATION AND
EMPLOYMENT GROWTH

The existence of RTP appears to have been a significant factor in locating in the region—for firms both inside and outside the park. We estimate that 47 percent of the R&D organizations in the park probably would not have located in the Raleigh-Durham area if RTP did not exist (see Figure 5-4). These percentage figures translate into an estimated direct employment growth for the region of about 18,900 jobs.[20]

For the sample of businesses in selected industries that located in the region but outside the Research Triangle Park after RTP was founded, we estimate that approximately 16 percent would not have located in the region if the park had not existed. This translates into an induced employment impact from RTP of about 1,240 jobs in the region's high-technology sector.[21] Approximately one-half of the 16 percent probably would not exist anywhere if the park had not been created (see Figure 5-5).

The relatively high percentage of firms whose decision to locate in the region was based on the existence of the park seems to be related to the absence of other strong locational pulls besides the universities and to the relative autonomy of park businesses from other businesses in the region. That is, in the absence of the park (and the universities) there would be insufficient reasons for a large corporation to locate an R&D branch plant in the Raleigh-Durham area. This contrasts with the Stanford area, where a well-developed set of networks among firms provides a strong locational pull for new companies. Also, in the cases of Stanford and Utah, many of the entrepreneurs were already living and working in the region and thus were most likely to start their own businesses there regardless of the existence of a research park.

The estimates of employment stimulated by RTP, described above, do not include firm and employment growth in the region brought about by the household income multiplier from the payroll for the 20,140 jobs in the region induced by the park (18,900 in the park and 1,240 outside the park). Nor do they include the employment impact from park organizations' purchase of inputs from local businesses.[22] We estimate that the number of jobs generated in the region due to the household income multiplier is about 25,500. The number of jobs generated by the local purchases of all businesses that would not be in the region except for the park is about 7,400. The total number of jobs in the region for which the Research Triangle Park was responsible—that is, that

Figure 5-4. *"If RTP Did Not Exist, Would You Have Located in the Region?"*

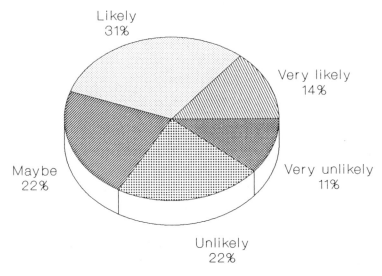

Likely
31%

Very likely
14%

Very unlikely
11%

Unlikely
22%

Maybe
22%

Percentage of R&D Organizations in the Park
(N ▪ 40)

Figure 5-5. *"If RTP Did Not Exist, Where Would You Be?"*

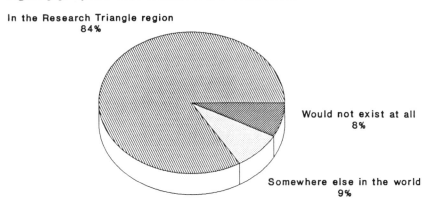

In the Research Triangle region
84%

Would not exist at all
8%

Somewhere else in the world
9%

Percentage of Out-of-Park High-Tech Businesses
(N ▪ 148)

would not be in the region if the park had not been created—is estimated to be about 52,000.[23] This represents 12.1 percent of the total regional employment in 1988 and 24.1 percent of the total increase in nongovernment employment since 1959, when the park was founded.

IMPACT ON INCOME GROWTH

Economic development should result in higher real income levels for a region. The per capita personal income of the Raleigh-Durham area grew from 93.0 in 1960 to 107.0 in 1987 as a percentage of U.S. per capita personal income. We cannot ascertain to what extent this increase was "caused" by the Research Triangle Park and its induced development because we do not have data on income levels of people working in the park or of people whose jobs were induced by the park. We do know that the proportion of professionals and managers of the park's workforce is substantially higher than that for the region as a whole, while the proportion of semiskilled and low-skilled jobs in the park's workforce is significantly lower than that for the region as a whole. If we assume that the wage or salary level of park and nonpark workers is the same or nearly the same as the national average for each occupational category, then there is little doubt that RTP has been responsible for a significant portion of the relative growth of the region's per capita personal income.[24]

IMPACT ON INCOME INEQUALITY

The impact of RTP on the degree of income inequality in the region cannot be estimated directly because of the absence of wage and salary data on park employees.[25] We do know, however, that the GINI coefficient of income inequality in the three-county Raleigh-Durham area decreased from 0.386 in 1960 to 0.353 in 1970 and then increased in the 1970s to 0.383 in 1980.[26] This trend is slightly below, and parallel to, the income inequality trend for the United States as a whole (see Figure 5-6). We also know from the occupational distribution of the workforce in the park that the earnings distribution is skewed to the high end but is not bimodal since there is only a small proportion of standardized manufacturing in the park. We therefore infer that RTP probably has not had any substantial effect on raising the level of earned income inequality of the region, given other factors.

Figure 5-6. *GINI Coefficients of Income Inequality in the Research Triangle Region*

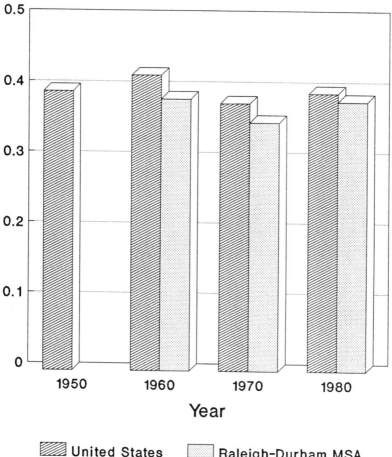

Year

United States ▦   Raleigh-Durham MSA ▨

Sources: U.S. Bureau of the Census, *1950, 1960, 1970, and 1980 Census of Population.*

IMPACT ON THE LOCAL LABOR MARKET

We expected that RTP has had a noticeable impact on the local labor market merely based on its size compared to the size of the region. Here, we examine the park's impact on the region's wage and salary structure, on the sources of labor supply to fill jobs created in the park, and on the participation of women and minorities in the workforce.

Evidence of the park's impact on the region's wage and salary structure comes from the perceptions of CEOs of organizations located inside and outside the park and from park management. There is agreement that RTP has been responsible for raising the salaries of certain classes of both professionals and nonprofessionals in the local labor market. The area's universities, in particular, have had difficulty in hiring and retaining key support personnel, including skilled lab technicians, programmers and data analysts, word processors, and administrative assistants. The impact on nonprofessionals' wages is partially due to the lack of depth of the local labor supply in a region that is not large and traditionally has not had a well-educated labor force excluding workers with university training. While the upward pressure on wage levels has been a problem for some area businesses, it has also helped mitigate the tendencies for increased income inequality discussed above.

A relatively large proportion of the park's professional workforce (48.3 percent) has been recruited from outside the region, including transfers from other branches (see Figure 5-7). Compared to those at Stanford and Utah, RTP organizations hire a somewhat larger proportion of their workforce directly from the area's universities, but they also hire a much larger proportion from branches of the same organization located outside the region. Businesses in the area supply a relatively low percentage of RTP's workforce, reflecting the underdevelopment of networks between park organizations and other businesses in the region.

The bulk of the park's nonprofessional workforce, on the other hand, comes from local sources, as one might expect. High schools and community colleges supply about 36 percent, while another 26 percent are hired from other businesses in the local labor market. Only 16.7 percent of nonprofessionals working in the park are from outside the region (see Figure 5-8).

Women comprise 45.5 percent of the park's workforce, while minorities comprise 15.0 percent.[27] In 1980, regionwide, women occupied 47.0 percent of the jobs (up from 38.6 percent in 1960), while minorities occupied 23.2 percent (down from 25.2 percent in 1960). We do not have data on the sex and race composition by occupational category of jobs in the park, so we cannot provide any evidence on the relative quality of jobs held by women and minorities. But we can say that women are about as well represented in the park's workforce as in the region as a whole. Minorities, on the other hand, are underrepresented in the park's workforce compared to the region as a whole.

To try to take into account some indirect effects of the park on women and

Figure 5-7. *Sources of Professional Workforce in the Research Triangle Park*

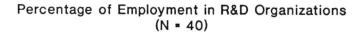

Other regional
businesses
16%

Area universities
29%

Other regional
sources
6%

Organizations' branches
not in region
26%

Other sources not
in region
22%

Percentage of Employment in R&D Organizations
(N ▪ 40)

Figure 5-8. *Sources of Nonprofessional Workforce in the Research Triangle Park*

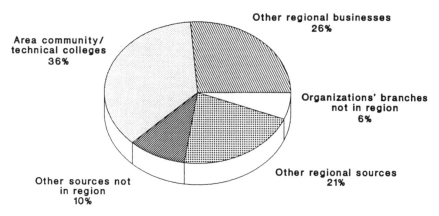

Other regional businesses
26%

Area community/
technical colleges
36%

Organizations' branches
not in region
6%

Other sources not
in region
10%

Other regional sources
21%

Percentage of Employment in R&D Organizations
(N ▪ 40)

minorities, we compared unemployment rates of these two groups between the Research Triangle area, the state, and the nation for selected years. In 1960, the average annual unemployment rate for women in the region was 70.1 percent of that for the state and 78.5 percent of that for the nation. By 1980, these figures were 63.2 percent and 61.4 percent, respectively. While certainly other events affected the relative regional unemployment rate of women for which we cannot control, the data at least suggest that the growth of RTP may have improved the overall position of women in the labor market. When the same calculations are repeated for minorities, the results show that in 1960 the minority unemployment rate in the region was 83.3 percent of that for the state and 71.0 percent of that for the nation. In 1980, these figures had declined to 67.0 and 57.2 percent, respectively. Thus, we can say that during the first twenty years of the park's operation, the position of minorities in the regional labor market improved substantially, although we cannot attribute this improvement solely to RTP.

IMPACT ON THE REGION'S INNOVATIVE CAPACITY

Without question, the innovative capacity of the Triangle region—that is, its ability to attract and generate innovative enterprises and individuals—has been positively affected by RTP. A region needs this capacity in order to adapt to changing external economic conditions, markets, prices, and policies and thus to sustain over time a high level of economic well-being. The Research Triangle Park has contributed to the development of the area's innovative capacity by helping to strengthen the research capability of the three research universities. Of particular importance in linking the three research universities to the park, both institutionally and through individual scientists, has been the Research Triangle Institute, now one of the largest contract research firms in the United States. Additionally, RTP indirectly has attracted innovative people to the region (who are not necessarily related to park organizations) by helping to place the Research Triangle "on the map" as an area that enjoys a substantial concentration of scientific talent and R&D activity, and by helping to create a special type of sociocultural milieu.

University administrators indicate that if the park had not been created, the overall quality of their university and its ability to generate sponsored research would either be somewhat lower or much lower. Among faculty members in selected departments of the region's three research universities,[28] about 20 percent of the sample respondents felt that the park had improved the quality and stature of their respective university considerably. Almost 70 percent of the faculty members perceived that the park had increased their graduating students' ability to get good jobs somewhat or considerably. RTP is also credited with having influenced (strongly or moderately) many respondents to accept the offer of a faculty position.

On the other hand, technology transfer from the park to other businesses in the region has not occurred to a large extent. Of our sample of technology-related businesses outside the park, only 13.6 of the respondents indicated that the park is a highly significant source of innovations or ideas. More specifically, only 7.7 percent said that unrelated park organizations are a very important or somewhat important source of R&D inputs to their business. The prestige and the economic vitality that RTP is perceived to bring to the region as a whole were much more frequently cited as important benefits of the park to other businesses (that is, as an external economy) than, say, as a source for purchasing goods or services or as a market for products.[29]

Until now, the incidence of spin-off businesses from RTP organizations has been low. Although we have no actual count of the number of spin-offs, in face-to-face interviews top managers from about one-quarter of the park organizations collectively could not cite more than three or four spin-offs from their organizations that have remained in the Triangle region. Many reasons have been offered to explain why the incidence of spin-off activity in the Research Triangle has not approached that found in Silicon Valley and Route 128: lack of venture capital, the organizational structure of establishments in the park, the lack of business service and support networks in the region, the absence of an entrepreneurial tradition, the restrictions universities place on faculty entrepreneurship, and the fairly stable economy that has kept (potential) entrepreneurs employed.[30] In our view, it is a combination of these factors, all of which are related to the relatively short time that any concentration of innovative and R&D activity (outside the universities) has existed in the region. We expect that the incidence of spin-offs will increase as the regional economy gains breadth and depth, and as the organizations located in the park become more linked to businesses outside the park.

IMPACT ON THE LOCAL POLITICAL CULTURE

Many local government leaders feel that RTP and its induced development have led to substantial changes in the political environment of the Triangle region. First, the sheer growth of the region as a result of the park has severely strained some of the region's infrastructure, especially roads and water and sewer facilities. This, in turn, has led to a number of local and neighborhood initiatives to institute growth controls ranging from the imposition of impact fees to moratoria on new building permits. Political debate in the region now centers around what is the desirable rate of growth and who should bear the cost of growth. Similarly, congestion on roads around the Research Triangle Park motivated the North Carolina General Assembly in August 1989 to authorize planning monies for a Regional Transportation Authority, which could lead to the first public bus service between the Triangle cities and RTP. It is doubtful that this set of issues would so dominate the political debate but for the growth of the park and related development.

Second, the Research Triangle Park and its induced development have helped to change the socioeconomic composition of the region dramatically. Because many park employees came from outside the region and even outside the South, the region has had to develop a new identity. One symbolic manifestation of this change occurred one weekend in 1983, when road signs proclaiming "Welcome to Durham, City of Tobacco" disappeared on Friday evening and were replaced by "Welcome to Durham, City of Medicine" by Monday morning's rush hour.

The change in the region's composition brought about by RTP has led to tensions between old-time residents and newcomers over such fundamental issues as investment in education, abortion, religion in schools, and environmental protection. These strains relate, in part, to the fact that the region that boasts of having the highest proportion of Ph.D.s and M.D.s of any metropolitan area in the United States also is home to a large concentration of residents, many born and raised in North Carolina, who have not graduated from high school.

Professionals working in the park have tended to become substantially involved in local civic organizations, chambers of commerce, and political organizations, though—with some exceptions—they have not sought public office. Businesses located in RTP, especially the larger ones, tend to command influence with state and local government leaders in getting their corporate, as well as their professional workforce's, needs met. Since much of the appeal of locating in the park or its environs has been the level of amenities, it is not surprising that these businesses have been highly supportive of measures to protect the environment, fund the public schools and state universities, and build additional transportation facilities to ease the congestion.

Perhaps the park has influenced local political outcomes most dramatically in the city of Durham. A large number of new residential areas sprang up south of the city in the 1970s and 1980s, mostly inhabited by newly located professionals and technicians working in the nearby park. Over time Durham annexed these areas, thereby significantly changing the racial composition and political coalitions in the city.

Finally, the location of the Research Triangle Park in the center of the triangle formed by Chapel Hill, Durham, and Raleigh and the resulting in-fill of the corridors between each of these cities and the park with residential, office, and retail developments have slowly led to a politics and set of policies promoting the *region* as the appropriate unit for planning, public facility investment, and service provision. Recent initiatives such as joint land use planning among local jurisdictions and the regional transportation authority are responses to the pressures placed by park-led growth on individual jurisdictions that were unable to cope with these pressures by acting autonomously. They also represent the desire of the modernizers, to whom Luebke and others refer, to build a "world-class region" that has a business, communication, physical, social, and public management infrastructure consonant with the world-class status of the Research Triangle Park and the three research universities.[31]

Conclusion

The Research Triangle Park, now thirty-two years old, has clearly had a significant economic impact on the region. Excluding the area's universities, the park itself represents about three-fourths of the high-tech employment in the area. Without RTP, almost two-thirds of the 32,000 jobs currently in the park would not be in the region. In addition, we estimate that the park has led to at least another 1,200 high-tech jobs in businesses that have located in the region but outside the park. Some 31,500 new jobs—both in new firms and in existing companies that have expanded—have been created in the region in response to the income multiplier from the aggregate payroll from the park or as a result of backward linkages from park organizations to local suppliers. The total regional employment growth that we attribute to the park is about 52,000, or 12.1 percent of the total number of jobs in the region in 1988.

Perhaps some of the less direct and less obvious impacts of RTP may be just as important. These include enhancement of the research capacity of the universities, helping to create a milieu that makes the region attractive for professionals who are not necessarily connected with the park, and providing the region visibility as a concentration of R&D and innovation that the universities by themselves probably would not be able to offer.

The park and the new business growth it has induced provide jobs disproportionately to professionals and managers. Clearly not all occupational or racial groups have benefited equally in the development of the park. Yet there is no evidence that low-income people and nonprofessionals in the region would be better off in the absence of the park. On the contrary, there is evidence that the wage levels of many nonprofessionals have increased as a result of additional labor demand from the park and the so-called wage-rollout effect.

The critical success factors for the Research Triangle Park seem to have been (1) the strength of the combined three research universities of the region in selected schools and departments, (2) the timing, or vintage, of the park, and (3) the vision, cohesiveness, and strong support of a key group of business, state government, and university leaders in understanding the need for a park and working for its success. While the universities were no match for Stanford or MIT, they collectively represented, arguably, the strongest concentration of academic talent in the South. The timing was critical because the South was about to take off in the long-term development of a modern industrial sector, and because in the early 1960s there was much less competition among a smaller number of areas competing for R&D facilities than there is now. The key role of individuals, from the genesis of the idea of the park to its entry into the maturation stage in the mid- to late 1970s, has been discussed above and is amply documented elsewhere.

The Research Triangle Park was conceived to stimulate the economic development of the state following an implicit development pole strategy (see Chapter 2). While the park has had a large-scale effect within the Triangle region itself, it has failed thus far to stimulate economic development, particularly manufacturing production facilities, in other parts of the state to the degree that it was intended and to the degree that development pole theory would predict. Even within the three-county Research Triangle area, the incidence of spin-offs from park organizations and of new, innovative-oriented business start-ups has been relatively small compared to other regions in the United States with large concentrations of high-technology businesses, although these types of impacts have started to pick up in the last two or three years.

We expect that the start-up rate of new, innovative, technology-based enterprises in the Research Triangle area will accelerate significantly in the next decade as the regional economy continues to mature through the development of networks of business and support services as well as sources of venture capital. It must be remembered that the business infrastructure to support new, technology-oriented business start-ups was virtually nonexistent thirty years ago when RTP was founded. In addition, though, the park management can help by changing some of its policies and priorities to create greater opportunities for small, highly innovative businesses to locate in the park, and to provide support services that these types of businesses typically need to succeed. Emphasizing locally owned, start-up businesses may also increase the potential for the location of associated manufacturing facilities in adjacent parts of the state.

The potential for the Research Triangle Park to achieve its original, overall objective of *statewide* economic development faces considerably larger challenges, however. Here, a large portion of the responsibility rests not on park management, but on the ability of state government and local school boards to improve the overall educational and skill levels of the nonuniversity-trained labor force of the state. The expectation that the manufacturing production facilities of high-tech corporations will be induced to locate in other parts of the state in order to have good access to the R&D activity in the Triangle region, a major premise for the creation of the Research Triangle Park, just will not materialize until the state has a labor force that is capable of operating and maintaining increasingly sophisticated equipment in the new knowledge-based economy of the 1990s and beyond.

6 The University of Utah Research Park

The University of Utah Research Park, opened in 1970, now occupies a 320-acre site adjacent to the University of Utah campus in the foothills overlooking Salt Lake City. As of 1988, the park contained fifty-seven organizations with a total workforce of approximately 4,200.[1] The park is a unit of the University of Utah. Policies are set by the university's administration and approved by the university's institutional council, which is chaired by the provost. This arrangement allows the park to be used as the key instrument in the university's promotion of "academic capitalism."

As we shall see, the case of the Utah Research Park represents an excellent example of the application of an indigenous development strategy. Rather than relying on the recruitment of branch plants of national or multinational corporations—which ends up transferring economic activity from other regions of the world—the University of Utah designed park policies to be responsive to the needs of local (largely faculty) entrepreneurs. As a result (and as an indication of the park's success), faculty entrepreneurs (and others) have been able to spin off new businesses from existing park organizations, a practice that has been responsible for much of the park's growth in the past ten years.

Historical Background

The Salt Lake City region now has a diversity of high-technology businesses that span a variety of industry sectors. Yet twenty years ago, the regional economy was highly dependent on the federal government (mostly for defense-related and NASA contracts) and the minerals and energy sectors. The volatility of these employment sectors and an ensuing brain drain from the state due to the narrowness of employment opportunities for university graduates were recognized by state government and business leaders as chronic problems. It became increasingly clear that the regional economy needed to become more diversified. In 1965 the U.S. Department of Defense designated federal land under its control near the university campus as "superfluous" and made it available to the state government. The university proposed that it be given the land for the purpose of building a research park.

The idea of creating a research park had its genesis with several prominent faculty members and administrators who had observed firsthand, as students or faculty members at Stanford, how a university could create an entrepreneurial milieu that later would stimulate high-tech growth in the larger region. The university's initial proposal to the governor for the excess federal land also prominently cited the brain drain that the state was experiencing, arguing that a research park could help generate a large number of job opportunities for scientists, engineers, and managers in the Salt Lake City region. In addition, the university had an acute need for "crunch space" in which to conduct several large contract research projects.

To provide additional support for its proposal, the university hired the consulting firm of Arthur D. Little, Inc., to conduct a feasibility study. Although the consultant's report was pessimistic about the chances of a research park succeeding, the university's president, James C. Fletcher (who shortly afterward was appointed director of NASA by President Richard M. Nixon), pursued the acquisition of the land and the authority to develop the research park. As a result of his efforts, the state legislature agreed to support the university's request for the land.

The university now had the land but did not have any funds to develop the necessary infrastructure—namely, water, sewer, and roads. It asked and subsequently received an enabling act from the legislature that gave the university broad authority over the park land and an exemption from local property taxes in exchange for "in-lieu of taxes" contributions to the city. In return for the anticipated stream of returns from the university's in-lieu contributions, Salt Lake City put in the necessary utilities and roads. If the university's negotiations with the state and city had not succeeded at this early and critical juncture, the park probably would not have gotten off the ground.

The first tenant to locate in the park, about two years after the park entity was formed, was the university itself under contract to the federal government. The federal government guaranteed to lease a new building owned by the university for ten years in order to implement a research contract with the university for the evaluation of the artificial heart. Thus, as in the case of many other successful parks, the federal government served as a principal early anchor of the Utah Research Park. At about the same time that the park was being formed, Dr. David C. Evans, a professor in the computer science department, and Dr. Ivan Sutherland, a colleague from MIT, formed a company that grew out of their university-based research on computer graphics. The Evans and Sutherland Computer Company moved into temporary building space on land adjacent to the park in anticipation of the park's development.

The second tenant, Terra Tek, Inc., was to have a significant influence on the future of the research park. Dr. Wayne Brown, chairman of the mechanical engineering department at the University of Utah, had previously established a very successful company, Kenway Engineering, based on contract research begun at the university for the U.S. Department of Defense and NASA. Encouraged by his success, Brown soon started a second firm, Terra Tek, as an outgrowth of his university-based research on rock mechanics. Brown located Terra Tek in the research park and developed a business philosophy of expanding the company's influence by spinning off viable groups as separate new technology-based companies. These spin-off companies were formed with equity shared by their own management, by Terra Tek, and by outside investors. Many of the spin-offs from Terra Tek also located in the park, including NPI, a world leader in plant biotechnology products.

Much of the subsequent growth of the Utah Research Park has stemmed either from spin-offs established by faculty members to extend contract research begun in the university or from spin-offs from companies already in the park. Indeed, Charles Evans, the park's director, estimates that over one-half of the current employment in the park can be traced to two entrepreneurs and former university faculty members—Wayne Brown and David Evans.[2]

Local Conditions and Resources

In 1970, the Salt Lake City region, consisting of Salt Lake and Davis counties, had a population of 557,600 and employment of 253,000. From 1970 to 1986, employment grew at an annual average rate of 5.4 percent (compared to a rate of 4.0 percent for the nation as a whole). By 1986, the metropolitan area had become the thirty-eighth largest in population in the United States.[3]

The Salt Lake City metropolitan area can be regarded as a peripheral area in terms of its access to markets, other metropolitan agglomerations, and leading research institutions (except for the University of Utah). It is 500 miles from

Denver, generally considered the regional center, and 850 miles from San Francisco. In between, there are no employment centers of any magnitude.[4] Salt Lake City can be considered a subregional center of the intermountain West. It contains a large number of regional corporate headquarters, and its airport serves as a hub to the northern Rocky Mountain and Pacific Northwest states. In 1987, the airport had an average of 204 commercial flights per day.

In 1970, the year the park was founded, the industrial structure of the Salt Lake City region was highly underrepresented in the manufacturing sector and overrepresented in mining, wholesale trade, and services. The area was underrepresented in high-technology industries, with only 5.0 percent of its employment in that sector compared to 9.8 percent for the United States.[5] The high-tech sector that was represented in the area was heavily concentrated in several large aerospace firms that were almost entirely dependent on defense and NASA contracts. A small, indigenous high-tech sector had developed but, again, it mostly depended on subcontracts from the large defense contractors.[6]

What the area did have was an entrepreneurially oriented research university that was particularly strong in the natural and health sciences and a well-educated and productive labor force. Now guided by the motto *academic capitalism*, the University of Utah traditionally has encouraged its faculty to engage in entrepreneurial activities while also fulfilling the normal university responsibilities of teaching and research. The university's attitude toward entrepreneurial activity seems to reflect the values and norms of the Mormon culture, which has a strong influence in the state. Over the years, the University of Utah has ranked high among all U.S. universities in the number of patented inventions coming from university-based research.[7]

Well above average educational attainment by Salt Lake City's labor force also reflects the importance that Mormon culture places on education. In 1970, adult residents of Salt Lake City had completed 12.6 median school years, compared to 12.1 years nationwide. The percentage of adult Salt Lake residents (twenty-five years or older) who had attended college was 31.9 compared to 21.3 percent for the United States, while the percentage that had not completed high school was only 31.5 percent compared to 47.7 percent for the nation.[8] As mentioned above, however, the lack of employment opportunities for highly skilled professionals had led to a brain drain from the region. Yet a good percentage of those who had left to take jobs elsewhere potentially could be drawn back to Utah by the pull of the Mormon culture, family ties, and the state's quality of life if job opportunities were available.

In summary, the principal factors that would enable the University of Utah Research Park to stimulate economic development in the region included (1) the entrepreneurial climate of the university, (2) the research strength of several of the university departments, (3) a supportive state and local government, and (4) an annual supply of well-trained graduates from the university and, more generally, a well-educated and productive labor force in the region. The factors mitigating against the park's ability to have a significant economic impact included the region's relative isolation from large metropolitan agglomerations and its low concentration of R&D activity outside the university and the aerospace industry. We also speculate that while the predominant Mormon culture supported an entrepreneurial climate in the region, it may have led to a hesitancy of some firms to locate there because of a perception that they might have difficulty in recruiting non-Mormons to move to the Salt Lake City region.

The Operations and Policies of the Park

When the research park was founded its most important objectives were (1) the diversification of the region's economic base, (2) the transfer of technology to other businesses, and (3) the creation of high-paying jobs in the labor market for local university graduates. Notably, the expansion of employment opportunities for lower-skilled workers was never considered as an objective of the park (see Table 6-1). Over time, it has become more important to develop and nurture new businesses, to enhance the prestige of the university, and to generate revenue from the leasing of park land and buildings.

Businesses that locate in the park lease land and facilities from park management. Generally, leases are for forty or fifty years, with rents based on appraised current market prices. Although average rent levels ($110,000 per acre per year) are not high by California or Route 128 standards, park management sees the rapid rise of rents in the research park and the lack of any "cheap" space as a disadvantage in its ability to continue its strategy of serving home-grown, start-up businesses.[9]

Responsibility for the provision of services is divided among park management, state and local government, and the tenants themselves. The principal role of park management is to plan and manage the physical environs of the park in order to maintain an attractive, campuslike setting for its tenants. Park management plans for the establishment of roads and utilities within the park,

Table 6-1. *Original Objectives for the University of Utah Research Park*

Very important
1. Diversify the region's economic base
2. Increase technology transfer to other businesses
3. Provide higher-paying jobs in the local labor market
4. Expand employment opportunities in the area
5. Increase employment opportunities for local university graduates

Moderately important
6. Develop and nurture new businesses
7. Capitalize on existing R&D in the area
8. Increase productivity of the economy through innovation
9. Encourage entrepreneurship in the region
10. Commercialize university-based research

Somewhat important
11. Enhance university's training capability through collaborative research
12. Enhance prestige of the affiliated university
13. Maximize profit from the development and lease of park land and buildings

Not important
14. Expand employment opportunities for low-skilled workers

Source: Charles Evans, University of Utah Research Park Director, interview with authors, July 8, 1988.

disposes of garbage, and serves as a liaison between park businesses and the University of Utah. It recently negotiated a contract to build a hotel with restaurants and meeting rooms inside the park.

State and local governments plan, finance, and maintain access roads to the park. The park is also served by local public transit (bus service). State government, along with the city and the university, assists park management with business recruiting. The city finances, constructs, and maintains roads within the park; it also pays for and provides water and sewer services and fire protection. The city and the university coprovide police protection/security. The tenants themselves are responsible for their individual business service needs (communications equipment, meeting rooms, consulting, and so on). They also must dispose of their own industrial wastes.

The permitted uses in the park include research and development, prototype manufacturing, light manufacturing, headquarters functions, and business support services. Non-R&D uses are permitted only if they are related to on-site R&D activity of the tenant organization. Park management has turned down offers of blue-chip corporations to locate facilities in the park because their activities would not have been accompanied by related R&D activity.[10] In marketing the park, the management has strong preferences for organizations that are R&D facilities, or at least are technology-oriented, and for locally owned and start-up businesses. Although light manufacturing is allowed, it is not encouraged.

The park has a private incubator facility for small, start-up companies that do not require a large amount of space. Currently, the incubator represents less than 3 percent of the total floor space in the park. Finally, the park does have a set of restrictive covenants that supersedes the R&D zoning classification. These restrictions do not include minimum lot sizes or a maximum footprint, however.

Characteristics of R&D Organizations in the Park [11]

A relatively large percentage of the organizations located in the Utah Research Park are single-plant firms (56.7 percent). Of those that are branches of multi-locational corporations, 61.5 percent have their headquarters in Utah. We can infer from these data and the reported recruitment policies of park management that the park has followed an indigenous or local development strategy.

The organizations in the park are significantly more oriented to product development and significantly less oriented to basic research than those in RTP or Stanford (see Figure 6-1). These differences are reflected in the occupational distribution of the park's workforce: compared to other parks, there is a smaller proportion of scientists and engineers (25.1 percent) and a larger proportion of skilled technicians (24.8 percent) (see Figure 6-2).

The technology area most heavily represented by the businesses in the park is biological and medical technology. A major reason for this specialty is the prominence of the University of Utah Medical Center, which has pioneered several well-known inventions including the artificial heart. Indeed, the Medical Center's need for crunch space was an early impetus for the park's growth. Other specializations are mineral- and energy-related technologies and computer applications (computer graphics, artificial language). In general, the park is not highly diversified in its technology areas. Rather, the areas reflect the particular strengths of the university. This specialization seems to be the result of self-selection of businesses rather than an explicit policy of the park.

Proximity to a research university, the preference of the CEO, the quality of air transportation service, and access to skilled labor have been the most important reasons for locating in the Salt Lake City region (see Table 6-2).

Figure 6-1. *Functional Specialization in the Utah Research Park*

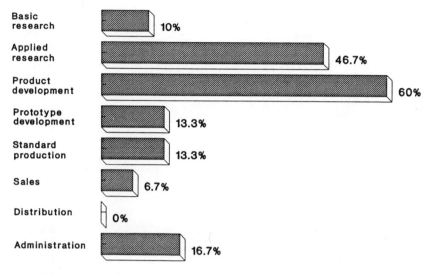

Functions

Basic research	10%
Applied research	46.7%
Product development	60%
Prototype development	13.3%
Standard production	13.3%
Sales	6.7%
Distribution	0%
Administration	16.7%

Percentage of Organizations with Function as First or Second Highest Effort
(N = 28)

Figure 6-2. *Occupational Mix in the Utah Research Park*

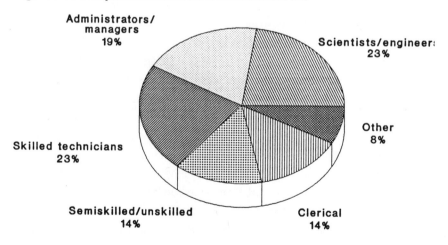

Administrators/
managers
19%

Scientists/engineer:
23%

Other
8%

Skilled technicians
23%

Semiskilled/unskilled
14%

Clerical
14%

Percentage of Employment in R&D Organizations
(N = 28)

Table 6-2. *Reasons for Locating in the Salt Lake City Region* (R&D Organizations in University of Utah Research Park, N = 28)

REASON	RANK IN IMPORTANCE
Proximity to research universities	1
Preference of CEO	2
Access to major airport	3
Access to skilled labor	4
Access to markets	5
Access to business services	6
Business climate	7
Other branches in the region	8
Access to unskilled and semi–skilled labor	9
Quality of public services	10
Concentration of firms in same or related industry	11
Cultural amenities	12
Physical climate	13
Quality and adequacy of infrastructure	14
Access to materials	15

Compared to the RTP and Stanford cases, linkages with other branches in the region, the preference of the CEO, and proximity to local markets are more important factors; the business and physical climate, cultural amenities, and the quality of the infrastructure are comparatively less important.

The advantages of locating in the research park versus other sites in the region are close proximity to the University of Utah and opportunities for sharing ideas and information with scientists, engineers, and managers in other technology-oriented businesses (see Figure 6-3). The suitability of the building sites in the park, land prices, and prestige of location in the park are much less important to organizations in the Utah park than to those in RTP and the Stanford park.

Most valued by park organizations about proximity to the University of Utah is access to faculty expertise and laboratories or other specialized facilities (see Figure 6-4). Access to the recruitment of university graduates is next most important. Compared to the RTP and Stanford cases, opportunities for employees to take university courses or special training and to take advantage of the cultural amenities associated with the university are less important potential

Figure 6-3. *Reasons for Locating in the Utah Research Park*

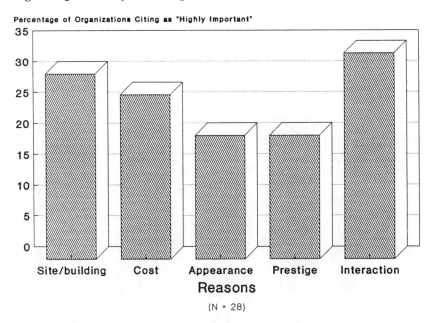

Percentage of Organizations Citing as "Highly Important"

(N = 28)

benefits of universities for organizations in this park. Likewise, while the literature suggests that access to university graduates for recruiting purposes is the most important reason for technology-oriented businesses to locate near a research university, this was not the case for the organizations in the Utah Research Park.

Park management cited several minor disadvantages of locating in the University of Utah Research Park versus other parks or regions. These include the high cost of leasing a site in the park, lower access to corporate headquarters and manufacturing facilities, inadequate supply of appropriate professional labor, and some restrictions on uses and design imposed by park management and the university.

Impacts of the Park on Regional Economic Development

We have attempted to estimate the direct and indirect impacts of the research park on various dimensions of regional economic development, including firm location, employment growth, personal income levels, income inequality, as-

Figure 6-4. *Benefits of the University to Utah Research Park Businesses*

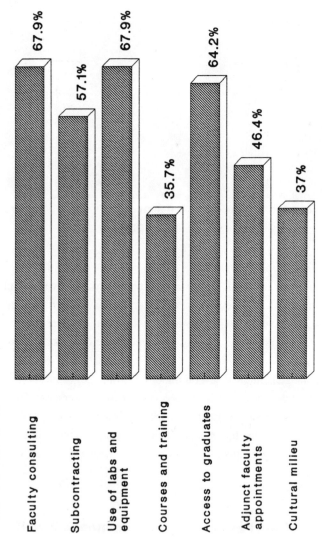

Potential Benefits

Faculty consulting — 67.9%

Subcontracting — 57.1%

Use of labs and equipment — 67.9%

Courses and training — 35.7%

Access to graduates — 64.2%

Adjunct faculty appointments — 46.4%

Cultural milieu — 37%

Percentage of Park Businesses Citing as Highly or Moderately Important
(N = 28)

Figure 6-5. *"If the Utah Research Park Did Not Exist, Where Would You Be?"*

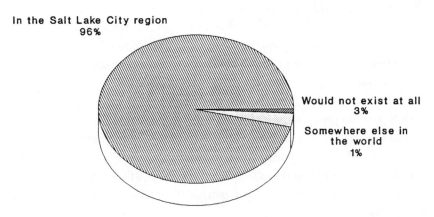

In the Salt Lake City region
96%

Would not exist at all
3%

Somewhere else in
the world
1%

Percentage of Out-of-Park High-Tech Businesses
(N ▪ 104)

pects of the local labor market, technological capacity of the region, and the political culture.[12]

I M P A C T   O N   F I R M   L O C A T I O N   A N D
E M P L O Y M E N T   G R O W T H

Of the organizations currently located in the park, 16.7 percent indicated that it is likely they would not have located in the region if the research park had not existed.[13] This group, which comprises 21.5 percent of the park's work-force, accounts for about eight hundred jobs. A similar question was asked of a sample of businesses that had first located in the region after the research park was created in 1970.[14] Of these, only 1.4 percent indicated that they would have located in some other region if the park had not existed, and 3.2 percent indicated that they would probably not exist anywhere but for the park. More than 95 percent stated that they would have located in the region anyway (see Figure 6-5).

From these data we infer that the research park itself probably has been responsible for bringing about one thousand high-tech jobs to the region, in addition to those in the park itself. Of these one thousand jobs, the net gain to the global economy that can be attributed to the Utah Research Park is approximately seven hundred.

The impacts of the park on firm and employment creation do not include the effects of increased consumer spending from the increased regional payroll (the household income multiplier) of the organizations whose location in the region was induced by the park. Nor do they include job growth due to local purchases of inputs by those businesses. We estimate that the household income multiplier has generated an additional 1,725 jobs in the region (in retail, consumer services, real estate, residential construction, and so forth), while local purchases of inputs to businesses have generated about 825 more jobs.[15] The total number of jobs in the region that we attribute to the Utah Research Park is 4,350. This represents about 1.8 percent of the total new jobs in the region since 1970.

There are no objective standards for judging whether the park-induced firm and employment growth in the region or the world is large or small. The numbers are small compared to the RTP and Stanford cases for several reasons. First, the Utah Research Park is small compared to the other two cases, so we can expect the total impact to be smaller. Second, because the Utah Research Park is newer, the process of induced firm location and the integration of park businesses with other local businesses through backward linkages have not had as much time to develop. Third, and probably most important, the park from the outset had a deliberate strategy of recruiting local entrepreneurs— specifically, University of Utah faculty and graduates—rather than businesses from outside the region. Most of these entrepreneurs would have started their businesses in the Salt Lake region anyway because of strong family or cultural ties. The same probably can be said for a relatively large percentage of new technology-oriented firms located outside the park.

IMPACT ON INCOME GROWTH

The growth of personal income in a region is another dimension of economic development that is not necessarily highly correlated with the rate of job growth. This may be particularly true in areas whose growth has been fueled by R&D activities and the arrival of high-tech businesses.

From 1960 to 1987, per capita personal income in the Salt Lake City region declined as a percentage of U.S. per capita personal income (see Figure 6-6).[16] The "after-park" change in relative per capita personal income (from 1970 on)

Figure 6-6. *Per Capita Personal Income in the Salt Lake City Region*

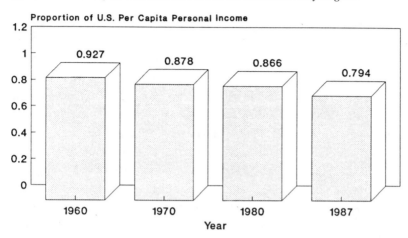

Sources: U.S. Department of Commerce, Bureau of Economic Analysis, *Local Area Personal Income* (tapes), and U.S. Bureau of the Census, *1960 Census of Population.*

was flat until 1980. Between 1980 and 1987, the region's relative per capita income level declined dramatically, in large part as a result of the depression in the energy sector. This decline illustrates that the R&D sector of the region's economy is still relatively small, and that successful high-tech development is not a panacea for all underlying problems in a given regional economy.

To control for broad regional factors—such as the effect of the drop in energy prices between the mid-1980s and 1990—and for the size of area on personal income growth, we compared the growth rate of regional total personal income for the ten-year periods before and after the research park was founded with the growth rate of total personal income for a control group of metropolitan areas of the same size class and in the same U.S. Census division for the same two periods. Between the 1960s and the 1970s, the Salt Lake City region experienced an increase in the personal income growth rate of almost six percentage points (from an average of 5.6 percent to 11.6 percent annually), while the control group experienced an increase of only 3.6 percentage points (from an average of 6.3 to 9.9 percent annually). Although this is evidence only of an association between park-led high-tech growth and income growth, it suggests that the high-tech–led development, of which the park was a part, was a factor in the income growth of the region.

IMPACT ON INCOME INEQUALITY

Some of the literature on high-tech development has asserted that high technology leads to increasing income inequality.[17] The argument is that high-tech businesses have bimodal wage and salary distributions, that is, a relatively large proportion of highly paid professionals but the rest lowly paid operatives. This does not seem to be the case with the economic development stimulated by the University of Utah Research Park. First, the bimodality appears to be much more characteristic of high-tech–production facilities—perhaps most vividly observed in microelectronics but also in instruments and pharmaceuticals. Businesses in research parks tend to have a smaller proportion of low-skilled or semiskilled workers because there is a much higher proportion of R&D activity and a much lower proportion of standardized production. In the Utah Research Park nearly one-half of the workforce consists of scientists, engineers, or managers and about one-quarter are skilled technicians (see Figure 6-2 above). Only 15 percent are semiskilled or unskilled. We also know from the data we collected on businesses drawn to the region by the research park or the university that these are likely to be other R&D facilities rather than production facilities. If the research park had induced a large amount of standardized production activity to the Salt Lake City region, then, ironically, we probably would be observing a greater amount of income inequality.

We collected U.S. Census data on the income distribution of households in the Salt Lake City region in 1970 and 1980 and calculated the GINI coefficient to measure the degree of income inequality (see Figure 6-7).[18] The GINI coefficient was also calculated for the United States as a whole for both years as reference norms. In 1970, the GINI coefficient for Salt Lake was just below that for the nation (0.3628 versus 0.3799). In 1980, the GINI coefficient for the region had increased very slightly to 0.3663, while for the United States it had increased by a significant amount (to 0.3943).

The two types of evidence suggest that while the research park was responsible for a modest increase in the number of relatively highly paid workers in the local labor market, it probably did not contribute to a bimodal distribution of income.

IMPACT ON THE LOCAL LABOR MARKET

According to local government and business leaders in the area as well as executives of high-tech businesses both inside and outside the research park, the

Figure 6-7. *GINI Coefficients of Income Inequality in the Salt Lake City Region*

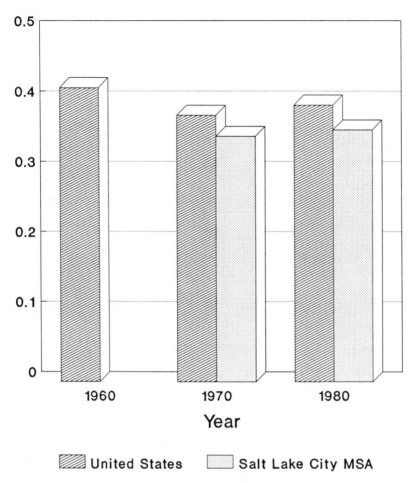

Sources: U.S. Bureau of the Census, *1950, 1960, 1970, and 1980 Census of Population.*

park has helped to push up the salaries of professionals by increasing the demand for this type of labor when the supply has not been able to adjust as quickly. On the other hand, there is an equally strong consensus among these individuals that the wage levels of nonprofessionals have probably not been affected in either direction by the economic development generated by the park (or the university). To the extent that this is true, the so-called industrial gentri-

Figure 6-8. *Sources of Professional Workforce in the Utah Research Park*

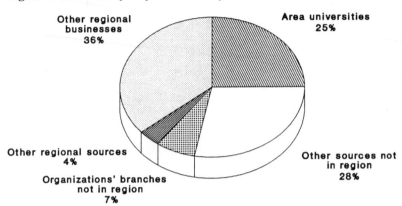

Percentage of Employment in R&D Organizations
(N = 28)

fication effect, where high-tech development has put upward pressure on wages that businesses in more competitive sectors cannot match and thus may go out of business or be forced to move, may not be a negative impact in this region. The down side of this development, of course, is that the relative wage levels of nonprofessionals in the area appear to have declined slightly.

Although scientists and engineers are usually thought of as being in a national labor market, a relatively high percentage of the professional workforce in the research park comes from local sources (see Figure 6-8). This is due to the park's indigenous development strategy—that is, many organizations in the park are spin-offs from other park businesses or from the university. We also speculate that the out-migration of professionals who are Mormon is lower than what one would otherwise expect at the same time that the in-migration of professionals who are not Mormon is also relatively low.

Just as one would expect, the sources of the park's nonprofessional work-force are even more local (see Figure 6-9). Compared to RTP and Stanford, a relatively large percentage of nonprofessionals come directly from area technical colleges or high schools. The percentage of nonprofessionals who have come from other businesses in the region is higher than in RTP but considerably lower than in Stanford, suggesting an intermediate level of integration of the park businesses within the local labor market.

The proportion of jobs held by women in the research park is less than the proportion of jobs occupied by women in the local labor market as a whole,

Figure 6-9. *Sources of Nonprofessional Workforce in the Utah Research Park*

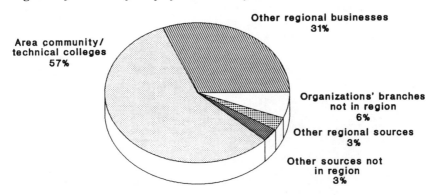

Other regional businesses
31%

Area community/
technical colleges
57%

Organizations' branches
not in region
6%

Other regional sources
3%

Other sources not
in region
3%

**Percentage of Employment in R&D Organizations
(N = 28)**

while the proportion of jobs held by minorities in the park exceeds that in the
local labor market (see Figure 6-10). Unfortunately, since we do not have information on the skill levels or wage/salary levels of these jobs, we cannot make
inferences about women's and minorities' share of the professional jobs in the
park.[19] We do know that the female unemployment rate in the Salt Lake City
region was 104 percent of the female unemployment rate in the United States
in 1970, while this percentage had declined to 83.0 in 1980.[20] We cannot attribute the improvement for women in the local labor market to the park since
we did not control for other possible causal factors. The evidence that we do
have suggests that the park probably has not improved the position of women in
the labor market directly, but that the economic development that the park and
the university have helped *induce* may have contributed to the improvement.[21]

IMPACT ON THE REGION'S INNOVATIVE CAPACITY

One potential benefit of a research park is that it can add to the innovative capacity of the region. This capacity refers to a region's ability to attract or spawn
innovative firms. It increases the potential of a region to adapt to changing
external economic conditions and to sustain its level of economic well-being.
A research park can contribute to a region's innovative capacity by increasing the research capacity of universities in the area, by serving as a source of
innovation for businesses outside the park, by enlarging the region's pool of

Figure 6-10. *Proportion of Jobs Held by Women and Minorities in the Utah Research Park*

Percentage of Employed Labor Force

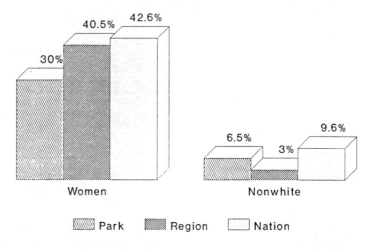

Sources: U.S. Bureau of the Census, *1980 Census of Population* (for regional and national data); responses to authors' questionnaires (N = 28) for park data.

entrepreneurs through spin-offs, and by serving as a symbol of innovation for the region.

There is a general consensus among University of Utah administrators that the research park has been of substantial benefit to the university. The university's ability to obtain sponsored research funds, to commercialize research, to draw and hold top faculty, and to attract high-quality graduate students is perceived to be either much higher or somewhat higher as a consequence of the research park. The overall quality and reputation of the university are perceived to be somewhat higher as a result of the park.

The research park has not served as an important source of innovation or R&D input to other technology-oriented businesses in the region. Nine percent of the businesses in our out-of-park sample said that the research park as a whole has been a highly significant source of innovations or ideas and only 4.3 percent believed that individual businesses in the park were important as sources of R&D. These data indicate that the park needs to be more integrated into the business community as a whole in order for the region to enjoy more fully the technology transfer benefits from the park.

The supply of highly skilled scientists, engineers, and entrepreneurs has been enhanced somewhat by the presence of the research park, but the University of Utah itself, rather than the park, probably has played a much larger role in this area. About 8 percent of the sample of out-of-park businesses cited the park as a highly significant source of labor supply, and 4 percent indicated that the park has been highly significant in making the region more attractive for workforce recruitment. But the percentage of businesses naming the university as an important source of labor supply was about 30 percent. The spawning of new entrepreneurs from existing park businesses has been, perhaps, the most important contribution of the park to the region's labor supply, at least if we believe that an entrepreneurial presence invigorates local economies.

Finally, the use of the research park as a symbol, or an image, to enhance a region's technological capacity should not be ignored. When asked what was most significant about the park for them, out-of-park businesses most frequently cited the prestige and sense of economic vitality that the park brings to the Salt Lake City region.

IMPACT ON THE LOCAL POLITICAL CULTURE

The final type of potential economic development impact (broadly defined) of the research park considered in this study relates to the local political culture. We are interested in knowing whether economic development stimulated by a concentration of R&D activity leads to changes in budgetary priorities, civic participation, and political/cultural tolerance. A small sample of local government, civic, and business leaders were interviewed about these relationships. There was a consensus that the presence of the research park, and the involvement of park employees in the community, had little or no bearing on local political outcomes and public expenditures for various categories. Respondents believed that, in general, park professionals' level of involvement in neighborhood or civic organizations varied considerably, although they rarely became involved in political organizations or were appointed to public positions (on commissions, boards, and so forth). The park and the development induced by the park were judged to have increased the sociocultural diversity of the community somewhat and to have increased the tolerance of the community toward sociocultural differences.

Conclusion

The University of Utah Research Park has been highly successful as an instrument of university policy to make its faculty more entrepreneurial. It also is an excellent example of an indigenous development strategy, since the park has been planned to meet the needs of local entrepreneurs (specifically, faculty members and graduates) rather than to attract branch plants of businesses headquartered elsewhere. Park planners hoped that the park would grow as a result of spin-offs from the university and local businesses, and to a considerable extent it has. More than half of all the current employment in the park can be traced to two former university faculty members, who spun off businesses to the park.

The success of the research park in stimulating the economic development of the region has been more mixed, however. We have estimated that about 4,350 jobs in the Salt Lake City region can be attributed to the park—that is, they would not exist if the park had not been created. This is a rather modest impact, in part because the park itself is not large, especially when compared to Stanford and RTP. A more important reason is that the park has not become well integrated with, or well linked to, the technology-oriented businesses in the region. It still remains an arm of the University of Utah, first and foremost. Much of the technology-oriented development that the area has experienced since 1970 probably would have occurred even if the research park had not existed. Twenty-eight percent of the high-tech businesses in our sample said that the presence of the University of Utah was a highly important or moderately important factor for locating in the region, while only 11 percent said that the presence of the research park was a highly or moderately important regional locational factor. It is difficult to disentangle the impact of the university from the research park completely. Yet it appears that the University of Utah itself has been more responsible both for creating a milieu that is conducive to stimulating local entrepreneurial activity and for bringing outside recognition to the Salt Lake City area as an "innovative" region.

There is no evidence that the research park has had any significant negative impacts, or that it has led to any sizable public sector opportunity costs by taking subsidies that could have been better used for other purposes. The park probably has not had a substantial impact on the position of women and minorities in the labor market in either direction because the park is small compared to the size of the local labor market. Yet the park and its induced development may have contributed at least marginally to the overall improvement of women's status in the local labor market.

When we look back at the development of the Utah Research Park, the critical success factors seem to lie with several key individuals in the university, the institutional stand taken by the university to support enthusiastically the entrepreneurial activities of its faculty, and a unique cultural milieu that attracted many Mormons back to Utah when the labor market position of professionals improved. These factors seem to be more important than more systematic or economic factors such as an advantageous location, industrial structure, availability of venture capital, and depth and breadth of a business infrastructure.

This raises the question of the replicability of the Utah model in creating an indigenous strategy for other regions. Our tentative conclusion is that the particular constellation of success factors in this case is unique, but that these tend to be substitutable. The indigenous model, or strategy, can work in places where a more conventional approach of recruiting the R&D branch plants of large, established corporations would probably not work. The latter strategy would probably not have worked in the case of Utah. On the other hand, the indigenous strategy has limits, particularly in areas that are not large metropolitan regions with highly developed agglomeration and urbanization economies and a large, untapped pool of potential entrepreneurial talent. It is more practical to employ a strategy that allows combinations of approaches to be used, depending on the stage of park development and the development of its milieu, and that takes full advantage of the region's strengths and resources.

Finally, we have learned that even highly successful research parks take a longer time to reach maturity and to lead to changes in a region's economic structure than previously thought. Although the University of Utah Research Park is old compared to the majority of parks currently in existence, it is still in the early stages of its life cycle. Parks (including the Utah Research Park) that have not reached the maturation stage are unlikely to have played a significant role in the economic restructuring of their regions' economies. Accordingly, many of the eventual economic development impacts of the University of Utah Research Park on the Salt Lake City region are uncertain.

7 The Stanford Research Park

The Stanford Research Park in Palo Alto, California, opened by Stanford University in 1951, is widely regarded as the "granddaddy" of research parks.[1] It was developed at a time when no models for research parks existed. It is not surprising, then, that the park we observe today is considerably different from what it was thirty years ago. The evolutionary path of the Stanford Research Park is a consequence of changing university needs, the burgeoning post–World War II West Coast technology sector, the research capability and entrepreneurial spirit at Stanford University, the growing environmental consciousness of the surrounding community, and the adaptability of park planners and managers.

What was used for cattle and horse grazing in the first half of this century became a location for a few newly established innovative businesses with close Stanford connections in the early to mid-1950s and then the site for a diverse set of business establishments performing light manufacturing, headquarters, R&D, and business service functions, predominantly in the electronics, aerospace, pharmaceutical, and chemical sectors. Several of the early tenants that are still in the park—most notably, Hewlett-Packard, Watkins-Johnson, and Varian Associates—are now major multinational corporations. Their success very much tells the story of the park's success, since these companies alone occupy one-third of the space in the park.

Today, fifty-nine businesses (thirty-one R&D organizations)—on eighty-one sites, since many businesses occupy more than one site—in the Stanford Research Park employ approximately 28,000 workers and annually expend more than $1.25 billion in payroll and benefits, make close to $15 million in property tax, sales tax, and net utility payments to the city of Palo Alto, and generate nearly $5 million in income for the university. The park itself sits on 660 acres of land donated by Leland Stanford in 1885 as part of his original endowment of 8,800 acres to establish Stanford University. Consequently, most park businesses are within one mile of campus. The park is just thirty-two miles south of San Francisco and within commuting distance of San Jose and Silicon Valley.

Historical Background [2]

The genesis of the Stanford Research Park (originally called the Stanford Industrial Park) can be related to two developments in the late 1940s. First, after World War II, when enrollment swelled from 4,500 to 8,500 students, Stanford University experienced a budget crunch. Faculty salaries were low, especially compared to the elite eastern universities with which the board of trustees wanted to compete, and there were no employee benefits (including health and retirement plans) for nonfaculty staff and administrators. Moreover, Stanford's endowment had not done particularly well between 1910 and 1950. In 1910, Stanford was the fourth best-endowed university in the United States, with financial assets of $18 million. In 1952, it had fallen to twelfth, with an endowment of $39.6 million, or only $13 million in 1910 dollars.[3] For this reason, the university aggressively sought ways to augment its endowment and increase its income.

Second, starting in the late 1940s there was a growing demand for business location sites near Stanford. At the time, Stanford's School of Engineering was graduating more engineers than MIT; a large number of these graduates (many of whom had come from the East Coast) sought jobs in the Palo Alto area, but because industrial location sites were scarce, the supply of engineering jobs near campus was not keeping pace with the growing demand. At the same time, newly formed companies, many with ties to the university, sought locations near Stanford to have access to the engineering graduates and faculty and to be able to send personnel to the university for advanced training.

It is important to note that key figures at Stanford welcomed close ties with local businesses. Perhaps the most notable of these was Frederick Terman, who had founded the university's Department of Electrical Engineering in the 1930s. After the war Terman returned to Stanford from Harvard with a plan to develop "steeples of excellence" in microwave engineering and solid state electronics,[4] for which close ties with industry were crucial. Stanford's president, Donald B. Tresidder (1942–48), was sympathetic to that goal, in part, because it was consistent with Leland Stanford's wish to "assist by experimentation and research in the advancement of useful knowledge and in the dissemination and practical application of same." Another indication of the university's desire to forge close university-industry ties was its creation of the Stanford Research Institute in 1946 for the purpose of "pursuing science for practical purposes [which] might not be fully compatible internally with the traditional roles of the university." Putting these two developments together, we can say that there were supply-side and demand-side reasons for establishing a research-oriented industrial park at Stanford in the first years of the 1950s.[5]

THE SUPPLY-SIDE MOTIVATION: INCOME

The members of Stanford's board of trustees and administration, including Alf Brandin, who had been hired as business manager in 1946, recognized that the university's undeveloped land was a potential source of revenue. At the time, much of it was rented out to cattle and horse ranchers who needed grazing land. Traditionalists on the board of trustees and among the faculty and administration were leery of the university acting as a developer.[6] There was a model at another West Coast university to draw from, however, for the University of Washington in Seattle recently had been given prime real estate—the Olympia Hotel and surrounding businesses—to manage.[7]

In the late 1940s, Stanford officials agreed that some of the university's undeveloped land (about 3,300 acres) had to be reserved for future academic and research needs. Brandin and a real estate adviser he had consulted, Colbert Coldwell (founder of Coldwell Banker), convinced the trustees that residential, other agricultural, commercial, or industrial uses might be appropriate for part of the remaining 4,500 acres of Stanford land. On further consideration, residential and agricultural uses were disregarded because of their limited ability to generate revenues.[8]

The university proceeded on two fronts. It pursued the development of a shopping center on Stanford property, and it explored the possibility of leasing sites to industrial businesses. The first piece of industrial land—fifty acres fronting El Camino Real, a main north-south transportation route—was designated for use in 1947.[9] Stanford required prospective tenants to abide by the LM (light manufacturing) zoning that the municipality had imposed, which established a minimum parcel size of one acre and "smokeless" operations. It also required them to submit plat and landscaping plans, building elevations, and a color scheme, which would become part of their leases. According to

Lowood, "some thought had been given to planting a line of trees to reduce [the site's] visibility from campus, but the potential size, nature, and identity of its occupants cannot be found on any documents of this period." In fact, says Brandin, Stanford planted a row of trees on the property to screen it from a railroad right-of-way.[10]

One fact was clear about these and subsequent parcels of industrial land: they could not be sold. In his endowment Leland Stanford had stipulated that the land "was to be held in trust and used to further the educational purposes of the university."[11] The inability to sell land, fee simple, meant that the university had to structure rental arrangements that would be attractive to tenants and at the same time satisfy the university's need for income.

THE DEMAND FOR LAND NEAR
STANFORD UNIVERSITY

Between 1940 and 1950, the metropolitan area in which Palo Alto is located (San Jose MSA) experienced a 66 percent increase in population. Population growth in the Palo Alto–Menlo Park part of the MSA was even more rapid. This growth heightened the demand for all types of land uses near Stanford University, including land for industrial use.

The development of new technology-oriented companies by individuals with a Stanford connection increased the demand for industrial sites near the university. One of these companies, Varian Associates, was formed to commercialize two technologies that were developed collaboratively with Stanford faculty: the klystron tube and nuclear induction, both of which were patented by the university.[12] Varian Associates first located in San Carlos, several miles north of Palo Alto. The company hired many Stanford faculty members as consultants and research associates and students as research assistants. By the late 1940s, Varian had outgrown the San Carlos site. Its solution was to move its administration and R&D operations closer to Stanford and keep an expanded manufacturing division in San Carlos: "This move would bring the company closer to old friends, ease ongoing collaborations, and improve access to graduate students in physics and electrical engineering."[13]

In 1949 or 1950, Russell Varian proposed to Alf Brandin that his company lease some of the acreage that the university recently had designated for industry. The proposal was embraced by the university, which agreed to a ninety-nine-year lease (with an inflation index) for ten acres, with an option for six more.[14] Varian's price was $4,500 per acre per year, which, according to Brandin, was "slightly more than the fee value of property already zoned for industrial use within one mile of the site." Varian Associates received a favorable government loan and began to build its new facility in September 1951. The firm assumed responsibility for all on-site development and design.[15]

Around the time Varian began construction, a second company, Eastman Kodak, leased fifteen acres for a photo-processing plant. Within three years, Stanford signed leases with, among others, General Electric; Houghton Mifflin; Scott, Foresman; Admiral Corporation; Preformed Line Products; Hewlett-Packard; and Beckman Instruments.[16] Of these, Hewlett-Packard became the largest tenant. The lineage of Hewlett-Packard is generally well known. Bill Hewlett and David Packard had been students of Frederick Terman. Initially, they located in South Palo Alto near a rail line but, like Varian, quickly outgrew that location. David Packard, who then was a Stanford trustee, asked Brandin for space in the research park.

To house these and others tenants, Stanford designated more and more acreage for industrial use. By 1955, the park's size was approximately 220 acres. In 1956, another 125 acres were added—mostly to accommodate Lockheed Corporation's Missiles and Space Division. By 1960, more than thirty companies held leases for industrial sites on the Stanford property. Expansion after 1960 became somewhat more difficult, primarily because it required the use of more environmentally sensitive land in the foothills. But the Palo Alto voters approved expansion in a 1960 referendum, and the park continued to grow. By 1977, it encompassed 652 acres. As new leases were signed, the university refined its locational requirements and lease terms. Thus, the evidence suggests that the Stanford Research Park developed organically rather than as part of a grand plan.

A related demand-side pressure came from the city of Palo Alto. Because its rapid growth was directly related to Stanford's expansion, some citizens felt that the university should annex more of its land to the city so that Palo Alto would receive more tax revenue.[17] Stanford's existing practice was to zone land and annex it to Palo Alto only before tenants took occupancy. Until that was done, taxes on the land used for nonacademic purposes (for example, for agriculture) were paid only to the county and at a lower rate.

ORIGIN OF THE "PARK" CONCEPT

According to Alf Brandin, Stanford's development site was the first to be referred to as an industrial "park." We find references to industrial parks elsewhere before 1947, but Brandin and others at Stanford may not have been aware of them.[18] Brandin's story is that in discussions about the use of Stanford property for industry in the late 1940s, the president of the Stanford Club of Denver referred to the parklike setting of residential neighborhoods in his town. He noted, in particular, the absence of fences. He suggested, says Brandin, that the land used by industry at Stanford could be regulated to appear similarly parklike in order to be compatible with the surrounding land uses (an academic campus and agriculture).

As mentioned above, the evidence we have reviewed suggests that the physical character of the park was shaped over the first decade or so of development, rather than as a grand plan in 1951. Many of the restrictions that accounted for the parklike milieu were set by the city of Palo Alto, sometimes at Stanford's request. The city's zoning for the park required a minimum of one acre per lot, a maximum building-to-ground ratio of 40 percent, and fifty-foot setbacks, and it proscribed heavy manufacturing. Tenants also were subject to signage, architectural, screening, and other restrictions that were imposed either by the city, Stanford, or both. These restrictions were ratcheted up over time as the price of park land rose. Today, in addition to lot size, footprint, floor area ratio, and setback and parking requirements, tenants in the park are subject to architectural review standards imposed by both the university and, since 1983, the city of Palo Alto. The Lands Management Department must approve new building plans proposed by existing and prospective tenants, and the plans must be submitted to Palo Alto's Architectural Review Board. There also are special setback requirements for properties near residential developments.

While relatively strict locational requirements were established from the beginning, it took several years for the university to work out enforcement problems. For example, when Kodak's architects submitted plans to the university for approval, they did not include details of an unsightly mechanical structure on the roof. After the building was completed, the university changed its review procedures for subsequent tenants. (Kodak later built a screen around the unsightly structure.)

In the early years, there was some debate at the university about the types of tenants it should accept and the size of the park relative to other uses. A Faculty

Advisory Committee on Land and Building Development, established in 1951, and a Master Plan for all university lands, submitted by Skidmore, Owings, and Merrill in 1953, focused on housing construction and retail use, with little discussion of light industry. In December 1952, *Business Week* reported that Stanford, while "looking for shiny-faced plants like laboratories and pharmaceutical manufacturers," also thought "an insurance company or two might make excellent tenants." [19] Some in the university community, most notably Frederick Terman, were concerned that the park not divert land from campus-related uses—for antennas and the linear accelerator, for example. Terman also urged the careful selection of tenants so that all park occupants would provide some benefit to the university.

By 1960, Stanford had developed a coherent approach to the use of its industrial park lands. A *Faculty-Staff Newsletter* in that year reasserted the trustees' intention to use part of Stanford's land for industrial purposes: "Initially, large parts of the development lands were to be devoted to single-family residences. But new factors have caused the University and its planners to turn more and more to consideration of light industrial development . . . a new business phenomenon has arisen in that companies wish a park-like and creative setting for their research and development people, and . . . location on Stanford land not only satisfies this company objective but also advances the University's academic program." [20] The year 1960 also marked the opening of the second stage of park development across Foothill Boulevard in the hills above the university. These 254 acres were annexed by Palo Alto following a referendum. The zoning approved as part of that vote was more restrictive than the earlier zoning.

Local Conditions and Resources

Students, alumni, and local residents often refer to Stanford University as "The Farm," harkening back to how its land was used before 1885. Using that metaphor, Stanford was a fertile farm on which to grow a research park in the early 1950s. Conditions in the park continued to be favorable for additional R&D growth in the 1960s, 1970s, and 1980s. Among these conditions were an abundance of Stanford-owned land with limited options for its use; a willingness by key actors at the university to use the land for light industry; an entrepreneurial spirit at Stanford that not only made close university-business relationships attractive, but also helped to generate new tenants desiring locations proximate to the university; overall growth in the Palo Alto area, which generally increased the demand for developable land; and a strong base at Stanford in engineering, electronics, and the physical and life sciences—fields from which inventions are likely. Moreover, business owners, skilled workers, and others have found the scenic beauty and temperate climate of the Palo Alto area and broader surrounds to their liking. Other important advantages provided by the park include access to a large, diversified, and well-educated pool of labor; the availability of business and financial and venture capital services; agglomeration economies; access to domestic and international markets; and location on the San Francisco Peninsula.

Officially, Palo Alto is in the San Jose metropolitan area. But it is almost as close to San Francisco as it is to San Jose. Those metropolitan areas together had more than 2.5 million inhabitants in 1950. Because of the size of the region, businesses locating in Palo Alto have had access to a large and diverse pool of labor.

Even in 1950, the labor force near Stanford was better educated than in many other places. Then, residents in the bicounty area completed approximately 11.5 median years of school, compared to 9.3 years nationally, and approximately 11 percent of the population attended four or more years of college, compared to 6.4 percent nationally. The area continued to contain relatively well-schooled workers in the following decades. The educational levels of local residents reflect the influence not only of Stanford but also of the University of Santa Clara (which has a reputable engineering program), San Jose State University, and two sizable community colleges in the region: De Anza and Foothill. Many other postsecondary institutions are located in San Francisco and the East Bay.

In addition to workers, Palo Alto businesses have had access to a wide variety of essential services, including legal, accounting, machinery repair, and financial. The last of these is especially important. The greater Bay area already was the West Coast's financial center, having in the Bank of America one of the nation's largest depository institutions. In a 1958 newsletter, American Trust Company and Banking boasts of having attracted $1.3 million in deposits within a month of opening its Stanford Industrial Park office in 1958 and details how those monies will be lent to help the park grow. In the 1970s and 1980s, additional funds for investment were made available by venture capitalists. A multitenant complex that opened near the park (3000 Sand Hill Road, in Menlo

Park) in the late 1970s attracted several venture capitalists. These companies have helped finance spin-offs in the research park and elsewhere in the region. The Sand Hill Road complex has been among the most active centers of venture capital in the United States, making available approximately $90 million per year for new undertakings.[21]

The creation of the research park provided the opportunity for similar businesses to locate near each other. As the park developed over the next several decades and as related businesses located in Silicon Valley to the east, the strength of this agglomeration grew.[22] In 1951, only about 5 percent of the bi-county workforce was engaged in high-technology or R&D–intensive industries (versus 9.4 percent for the United States and 10.1 percent for California). By 1986, that percentage had grown to over 20 percent (compared to 8.5 percent for the United States and 11.7 percent for California).[23]

In addition to its strategic location in the San Francisco Bay area, the Stanford Research Park is situated 150 miles from the state capital in Sacramento and 425 miles from Los Angeles. Palo Alto is served by the Southern Pacific Railroad and CALTRANS, which runs along the San Francisco Peninsula, and by major airports in San Francisco and San Jose. Today, those airports have over 650 flights per day.

In the early 1950s, Stanford's West Coast location was strategic because of the growing economy of the Pacific states. Today, it is strategic because of the increasing importance of the Pacific Rim countries. Many of these countries have become the home of the semiconductor and electronic products manufacturing and assembly plants that initially located in Silicon Valley.

The Operations and Policies of the Park

The Stanford Research Park is managed by the Lands Management Department, an administrative unit of the university. The director of the Lands Management office reports to the vice-president for administration. General policies for the park are set by the university's board of trustees.

ZONING AND LAND USE DECISIONS

The modus operandi for development is for Stanford University to improve the park land (that is, build access roads, sewer and water hookups, and underground telephone and electricity connections) and then petition Palo Alto to annex the parcel(s).[24] By agreeing to annex the land, the city assumes responsibility for maintaining the infrastructure and providing municipal services, and it gains the right to levy property taxes and collect a one-cent sales tax. In order for the municipality to annex the property, either the land must be developed in a way that is consistent with the zoning map (which includes the Stanford land) or the zoning must be changed.

In the late 1940s, most of the land that is now in the park was zoned for agriculture. The university asked Santa Clara County to change the zoning to light manufacturing, or LM. (At that time, the city of Palo Alto had no planning staff.) The land developed from 1951 to 1960 was within the area initially zoned for LM. That classification requires minimum one-acre building sites, a 30 percent maximum footprint, a 40 percent floor area ratio, and one car space per three hundred square feet of building space.

During the 1951–60 period, there was at least two zoning controversies. In 1956, Stanford asked the city to rezone part of the park from LM to P-C, or planned community, to allow the construction of a gas station and a bank in the park. The City Council, under pressure from local business people, originally defeated the proposal. Later, the Planning Commission rezoned two parcels of land for a gas station and bank rather than changing the park's overall zoning.[25] Also during this period residents of the foothills tried unsuccessfully to incorporate into a separate municipality to protect the surrounding land.[26]

In the late 1950s, Ampex Corporation, a manufacturer of precision recording equipment, sought a 31-acre site in the Los Altos hills. To accommodate Ampex, Stanford offered a new 254-acre tract to Palo Alto for annexation. The City Council approved the annexation and zoned the area LM-5, requiring minimum five-acre lots, a maximum 15 percent footprint, one hundred–foot setbacks, a 30 percent floor area ratio, and a 35 percent open space set-aside.[27] Concerned about the environmental impact of developing the foothills, a group of citizens had the issue put to a referendum. The university lobbied vigorously for the initiative, and it passed in November 1960.

Opposition to foothill development continued in the 1960s and 1970s. In 1972, the Committee for Green Foothills sued the university and Xerox Corporation to prevent Xerox from locating in the foothills. The university resolved the matter by agreeing to put four parcels of land adjacent to the Xerox site into long-term contracts to be maintained as open space.[28] In the late 1970s, the Mid-Peninsula Regional Park District (later renamed the Regional Open Space

District [ROSD]) and the Peninsula Open Space Trust were formed. Using the power of eminent domain, the ROSD and the trust buy access easements and land in fee simple from property owners in the foothills to provide public open space. Twenty thousand acres already have been condemned and purchased for maintenance as open space but none yet from Stanford.

The most recent land use issue arose in the early 1980s with the discovery of at least ten toxic waste sites in the research park, caused in most cases by leaking storage tanks. The U.S. Environmental Protection Agency requires these sites to be cleaned up. So far, the tenants responsible for the problem have borne most of the cleanup costs, though Stanford, as landowner, is also legally responsible. These hazardous waste problems have led the city to add strict containment and disposal regulations to its planning requirements.[29]

LEASING ARRANGEMENTS

Tenants lease land from Stanford under a variety of terms. In most cases, the tenants construct their own buildings, which on completion technically revert to the university. The university is given the right of first refusal for leasing or buying facilities if a tenant has to terminate before the lease period ends. Stanford is not required to renew leases that expire, though nonrenewal would be unusual. According to Zera Murphy, the park director, "each renewal request is considered individually."[30] Because most of the sites in the park (those leased before 1978) are for long periods at a fixed rent, many of them are rented by intermediaries who profit from the spread between the fixed rent they pay and the market rent they can charge the sublessees to whom they rent.

As noted above, the terms of the first lease—with Varian Associates—were occupancy for ninety-nine years (with an inflation index) and rental at $4,500 per acre per year. In the late 1940s and early 1950s, there had been considerable discussion among Stanford officials about lease terms that would be both favorable to the university and attractive to tenants. Because many tenants preferred to buy land fee simple and because leaseholds were not used elsewhere in California, a long lease was judged to be necessary to ensure demand. A long term also was deemed necessary so that tenants could secure financing for construction. California state law stipulated that any term longer than ninety-nine years would be tantamount to a sale of land, so ninety-nine years was the longest lease that was consistent with Senator Leland Stanford's proscription on land sales.

Twenty-two leases following Varian's, until 1960, were also for ninety-nine years but were prepaid. (Ten prepaid leases were for other terms, ranging from thirty-seven to ninety years.) Stanford favored prepaid leases because it did not have to pay taxes on the up-front payment. The lump-sum payments also gave the university immediate income with which to finance improvements in the park and make equity investments that would provide a hedge against inflation.

Between 1960 and 1978, most lease terms were shortened to fifty-one years but were still prepaid. (Those fifty-one-year prepaid leases account for approximately 60 percent of all current leases.) Fifty-one years was chosen as the term because most buildings could be amortized in fifty years, plus having one year for construction.[31] In 1978, Stanford reintroduced annual rentals. At present, land typically is reappraised every five years (negotiations with tenants sometimes lead to a different interval). Until the reappraisal, rents are adjusted for inflation. If land values dropped over a five-year period, rents would not, according to Zera Murphy.[32] In 1988, the average value of leased space in the park was $45 per square foot. At that rate, the 9 million square feet of leased space produces gross revenues of approximately $5 million per year.

With annual leases and inflation adjustments, Stanford is able to set rents to achieve a competitive rate of return on its land assets. Most recently, this rate of return has been 10–11 percent. In earlier years, when rents could not be adjusted annually, we suspect that the rate of return may have been less than the competitive rate. Alf Brandin admits that "Stanford could not command as high a price [in the early years] as it later could to reflect the attractive design and amenities."[33] He also notes that university officials did not expect land prices to increase as rapidly and as soon as they did. In June 1988, the board of trustees approved for the first time explicit criteria for Lands Management to apply when considering lease extensions.

THE PROVISION OF SERVICES AND AMENITIES

Infrastructure, utilities, and public services to tenants in the research park are provided by the university, city and county, state, and private sector. In the early stages of park development, Stanford shared the cost—with Palo Alto and Santa Clara County—of sewer lines and improvements to Page Mill Road, a major road that traverses the park. The university helped lobby the state department of transportation for improvements to state roads, including El Camino Real. The state financed the construction of I-280 that serves the park from the east, and the university helped the California Department of Transportation lay out the routing of the interstate in the 1950s (seven miles of I-280 traverses Stanford property).[34] Other roads in the park were built by the university without public monies (for example, Hansen Way).

The regional transit systems, managed by the Santa Clara and San Mateo counties' transit authorities, provide public bus transportation between the Stanford Research Park and neighboring communities. The cost of that service is borne, in part, by riders, with subsidies from the federal and state governments. Commuter train service is available from Palo Alto to San Francisco, with costs spread similarly.

Tenants receive water and electric services from municipally owned utilities and pay the city directly for their consumption. They also typically pay for water, gas, and sewer connections, though that is negotiated with the university on a lease-by-lease basis. According to William Zaner, the current city manager, Palo Alto receives more in taxes and user fees from park tenants than the city spends to provide services—not only water and electricity but police and fire protection as well. Both the water and electric utilities offer efficient service at a relatively low cost per unit. Electric power is particularly inexpensive because the city obtains power in bulk under favorable terms and distributes the power itself. Water is relatively inexpensive because it is accessible: seventy-two-inch lines from Hetch Hetchy Reservoir traverse Stanford University land. According to Zera Murphy, most amenities, including recreation and landscaping, are the tenants' responsibility.[35]

Characteristics of R&D Organizations in the Park

The fact that 55.5 percent of our sample of twenty-four businesses first leased land in the Stanford Research Park prior to 1970 underscores its maturity. Approximately 25 percent of the businesses moved to the park in the 1970s, and the remainder signed their leases in the 1980s.[36]

Only 9 percent of park businesses are single-plant firms. That is considerably less than for the Research Triangle Park (29.3 percent) and the University of Utah Research Park (56.7 percent). Of the 91 percent that are part of multifacility organizations, more than half serve as headquarters and another 20 percent have headquarters elsewhere in California. These percentages are substantially higher than for the other two case study parks. The relatively large

Figure 7-1. *Functional Specialization in the Stanford Research Park*

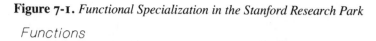

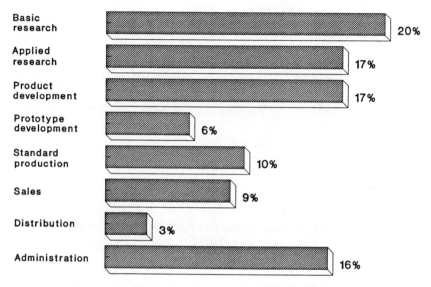

Percentage of Organizations with Function as Primary Activity
(N = 24)

number of headquarters and multifacility firms in the Stanford Research Park also demonstrates the park's maturity and the fact that it contains homegrown businesses that have expanded.

Of the functions performed in the park by tenants, basic research was cited most frequently as most important to the respondents' operations. That was followed in frequency by applied research and product development, administration, and routine production.[37] The importance of basic and applied research is an indication of the close university connection described below. (See Figure 7-1.) In comparison, organizations in the Research Triangle Park and the University of Utah Research Park are more oriented to applied research and product development.

The distribution of occupations in the park is shown in Figure 7-2. The Stanford park has a smaller proportion of scientists and engineers than RTP but larger than the Utah Research Park, and a larger proportion of administrators and managers than the other two parks as a result of more headquarters activity.

Forty percent of the park's employees are women, compared to 43.2 percent in Santa Clara County, 42.8 percent in California, and 42.6 percent in the

Figure 7-2. *Occupational Mix in the Stanford Research Park*

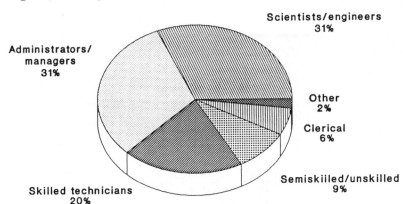

Scientists/engineers
31%

Administrators/
managers
31%

Other
2%

Clerical
6%

Semiskiiled/unskilled
9%

Skilled technicians
20%

**Percentage of Employment in R&D Organizations
(N = 24)**

United States. Twenty percent of the park's workforce comes from minority groups, compared to 11.8 percent for the county, 13 percent for the state, and 9.6 percent for the nation. The proportion of women is higher than in the Utah park but lower than in the Research Triangle Park. The proportion of minorities is higher than in the other two parks.

Approximately half of the park's professional workers (scientists, engineers, administrators, and managers) were hired from sources outside the Palo Alto area (see Figure 7-3). That is a considerably higher percentage than for the other two parks. Other sources of professional workers include area universities (19.5 percent) and other firms in the region (17.1 percent). Almost half of the nonprofessional workforce, including technicians, semiskilled and unskilled workers, and clerical employees, were hired from other regional businesses, a considerably higher proportion than in the other case study parks (see Figure 7-4). Of this group, 25.4 percent came from other sources in the region; 15.6 percent, from area community and technical colleges and high schools; and 5.8 percent, from other branches of the respondents' businesses. Only 4.5 percent were imported from sources outside the region. The proportion coming from area secondary and postsecondary institutions is much lower than in the Research Triangle and University of Utah parks, perhaps reflecting the park's need for more experienced personnel or an inadequate throughput of students

Figure 7-3. *Sources of Professional Workforce in the Stanford Research Park*

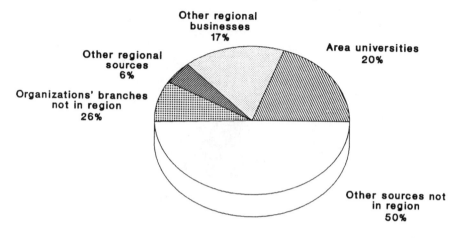

Percentage of Employment in R&D Organizations
(N = 24)

Figure 7-4. *Sources of Nonprofessional Workforce in the Stanford Research Park*

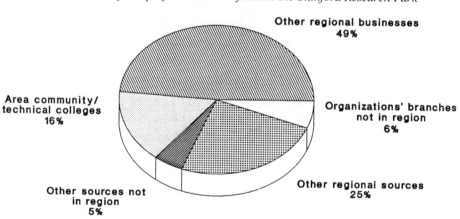

Percentage of Employment in R&D Organizations
(N = 24)

Figure 7-5. *Sources of Professional Workforce for Regional Businesses outside the Stanford Research Park*

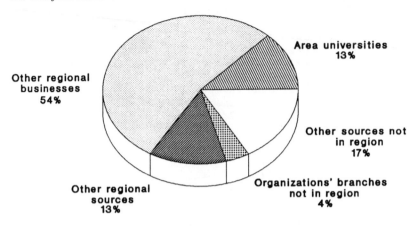

Area universities
13%

Other regional
businesses
54%

Other sources not
in region
17%

Other regional
sources
13%

Organizations' branches
not in region
4%

**Percentage of All Responses
(N = 89)**

from area schools.[38] For a breakdown of the sources of the professional and nonprofessional workforce of regional businesses outside the park, see Figures 7-5 and 7-6.

Finally, businesses in the Stanford Research Park, as in the Research Triangle and University of Utah parks, purchase a considerable amount of nonlabor inputs. Our best estimate is that Stanford park businesses spend approximately equal amounts each year on nonlabor and labor inputs, or approximately $1.1 billion in the aggregate. Table 7-1 shows the distribution of those purchases within the region, the state, the country, and the world. A greater proportion of nonlabor inputs are purchased from in-state sources by Stanford park businesses than by organizations in the Research Triangle and University of Utah parks, indicating the greater self-sufficiency of the California economy.

When asked to indicate the importance of various factors as determinants of its tenants' location in the region and the research park, park management claimed that all factors are equally important. Park businesses, on the other hand, felt that some factors were more significant than others (see Tables 7-2 to 7-4). For this group, the two most important determinants of location in the region were proximity to skilled labor and to Stanford University. Since much of the skilled labor comes from Stanford, these factors are related. This serves as further evidence of the importance of the university to research park

Figure 7-6. *Sources of Nonprofessional Workforce for Regional Businesses outside the Stanford Research Park*

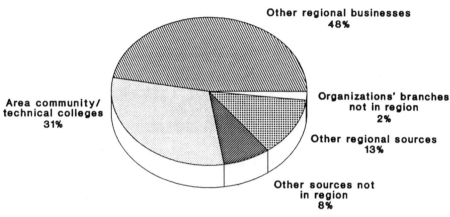

Percentage of All Responses
(N = 89)

businesses. (Businesses in the Research Triangle and University of Utah parks also ranked proximity to the university and to skilled labor near the top.) The third most important factor given was business climate, or the positive reputation of the region among business people. This reputation comes from various sources, including the very presence of a successful research park. The fourth location determinant cited by businesses in the Stanford Research Park (and the third most important factor mentioned by businesses in the other parks) is proximity to a major airport. This indicates the importance of access to other corporate facilities and centers of knowledge.

The most important reasons given for choosing a site in the park were the suitability of the land and building, the leasing arrangements, and the appearance of the park. To determine the importance of these qualities to the site location decision, we asked whether the businesses would have moved to the region if no park existed. Twenty-five of twenty-six respondents said they would have located in the region in any case.

The importance of Stanford University (see Table 7-2) is also revealed in responses to a direct question about why park businesses value proximity to the university. The two reasons mentioned most frequently (and equally) were access to technically trained graduates for employment and access to faculty for consulting. The next most frequent response was "courses and training for

Table 7-1. *Source of Nonlabor Inputs of Businesses in the Stanford Research Park (N = 24)*

% PURCHASED:	0–20%	21–40%	41–60%	> 60%
Within 20 miles of park	7	3	6	4
More than 20 miles from park but in California	8	9	2	1
Outside California but in U.S.	12	3	3	2
Outside U.S.	19	0	1	0

Table 7-2. *Reasons of Park Organizations for Locating in the Palo Alto Area* (R&D Organizations in Stanford Research Park, N = 26)

REASON	RANK IN IMPORTANCE
Access to skilled labor	1
Proximity to research university	2
Preference of CEO	3
Business climate	4
Access to major airports	5
Access to business services	6
Quality and adequacy of infrastructure	7
Physical climate	8
Access to unskilled and semi-skilled labor	9
Concentration of firms in same or related industry	10
Cultural amenities	11
Access to materials	12
Access to markets	13
Quality of public services	14
Other branches in the region	15

Table 7-3. *Reasons for Locating in the Stanford Research Park* (N = 24)

REASON	HIGH OR MODERATE IMPORTANCE (% of Respondents)
Land and building suitability	87.5
Leasing arrangements	66.7
Appearance of park	64.0
Prestige of park	60.0
Access to ideas/creativity	50.0

employees." These answers conform to the literature on the role of universities in economic development.[39] Twenty of twenty-six respondents had formal or informal ties to Stanford. These ties to the university were stronger than the ties between park businesses, since twenty-three out of twenty-four businesses had no, or only occasional, professional interaction with other establishments in the park.

Table 7-4. *Importance of Stanford University to Park Organizations* (N = 19)

FACTOR	HIGH OR MODERATE IMPORTANCE (% of Respondents)
Access to faculty consulting	83.3
Access to entry–level labor force	73.7
Opportunities to subcontract	66.7
Courses and training for employees	57.9
Cultural milieu	57.1
Faculty appointments for staff	44.4
Use of university facilities	38.9

More than most other parks in the United States, and certainly more than the Research Triangle Park, the Stanford park has close relations with the administration and faculty of a university.[40] There are two major reasons for this closeness; Stanford University administers the park and its faculty generally recognizes the benefits of park affiliation. Currently, between fifteen and twenty scientists employed by firms in the research park hold "courtesy faculty appointments" in academic departments. At the same time, as many as 15 percent of Stanford's faculty (or two hundred scholars) consult each year with park businesses.[41] Park businesses and related academic departments also jointly sponsor research seminars and colloquia. Occasionally, park management consults with university faculty on development issues and sometimes enlists faculty help to recruit prospective tenants.

Any visitor to Stanford University is struck by the number of facilities on campus that have been donated by firms in the park. The university development office and other administration officials stated that park firms also have given considerable amounts of unrestricted monies to the university but did not have figures readily available. In several areas of research (on semiconductors, for example), Stanford faculty use equipment owned by park businesses. University officials also estimated that as many as five hundred graduate students each year receive funds from park organizations for fellowships, part-time employment, and internships.

Impacts of the Park on Regional Economic Development

This section assesses the impacts of the park on businesses outside the park, the location of economic activity and job growth in the region, the generation of income in the region, income inequality, the distribution of economic outcomes among women and minorities, other labor market outcomes, and the region's innovative capacity, business climate, and political culture.

IMPACT ON BUSINESSES OUTSIDE THE PARK

Businesses outside the research park differ from park businesses in two important respects. First, they engage more heavily in routine production, as opposed to research and development, and thus employ a larger share of semiskilled and unskilled production workers. Second, many of the businesses in this group are located in Silicon Valley. That region of the state, which lies in Santa Clara County, contains many large electronic components assembly plants that employ a relatively large number of female and nonwhite workers. The modal distance of these businesses to the research park is twenty miles; 65 percent are between fifteen and thirty miles away. Relatively few of the nonlabor inputs that these businesses purchase come from other branches of their corporation, which is not surprising since most are single-plant firms. The large share of purchases are made within a thirty-mile radius of the plant, however.

Table 7-5 summarizes out-of-park businesses' reasons for locating in the Palo Alto area. The most important reasons include the preference of the CEO; proximity to customers, business services, and trained labor; and the overall business climate. These responses generally are consistent with those of out-of-park businesses near the Research Triangle and University of Utah parks. In contrast with the more heavily R&D firms in Stanford and the other parks, out-of-park respondents did not as frequently cite proximity to a research university or airport as important. Approximately 70 percent of the respondents said they had no formal or informal ties with Stanford University. (Recall that almost 80 percent of businesses inside the park stated that they had ties to the university.)

The mere existence of the Stanford Research Park has not induced many businesses to locate in the region. However, as Table 7-6 indicates, a larger proportion of respondents judged the park to be a benefit to their businesses. Table 7-6 also lists some possible benefits that out-of-park firms generally do not recognize or care about, including the provision of labor, ideas, and other inputs.

Table 7-5. *Reasons of Out-of-Park Businesses for Locating in the Palo Alto Area* (N = 86)

	% of Respondents	
REASON	HIGH/MODERATE IMPORTANCE	MINOR/NO IMPORTANCE
Preference of CEO	78.7	20.2
Access to markets	68.6	27.0
Access to business services	66.2	31.4
Access to skilled labor	60.4	35.9
Business climate	59.5	37.1
Access to materials	56.6	37.1
Concentration of firms in same or related industry	55.0	41.5
Quality and adequacy of infrastructure	52.8	46.0
Physical climate	49.5	46.0
Access to unskilled and semi–skilled labor	43.9	53.9
Access to major airports	36.0	59.6
Cultural amenities	36.0	59.6
Quality of public services	22.4	73.0
Proximity to research university	20.2	75.3
Presence of research park	11.2	88.8
Other branches in the region	5.6	86.5

Table 7-6. *Perceived Benefits of the Stanford Research Park to Out-of-Park Businesses*

BENEFIT OF PARK	SIGNIFICANT BENEFIT	SOMEWHAT OF A BENEFIT	NOT A CONSIDERATION
Creates economic prosperity/vitality	13.5	30.3	51.7
Brings prestige to the region	10.1	29.2	56.2
Serves as supplier of labor	5.6	19.1	70.8
Helps in labor recruiting	4.5	22.5	67.4
Serves as source of ideas, R&D	3.4	22.5	69.7
Serves as source of other inputs	1.1	16.9	77.5

To pursue further the relationship between the park and business location in the region, we asked respondents whether their organizations would be located in San Mateo or Santa Clara counties if the Stanford Research Park did not exist. Of our sample, 86.5 percent said that location in one of these counties without the park was likely or very likely, 5.6 percent said "maybe," and 5.6 percent said that location was unlikely or very unlikely. When we asked whether the business would exist anywhere if the park had not been created, the distribution of responses was similar. Businesses in the University of Utah Research Park region answered these questions in a similar way, whereas relatively more businesses in the Triangle region of North Carolina indicated that they might not have located there if the Research Triangle Park did not exist. That is not surprising since the smaller Triangle region, unlike the San Francisco Bay area, has no other major business agglomeration.

Businesses outside the park also were asked to specify the types of interactions they had with firms inside the park. Approximately 33.3 percent of the respondents sell supplies to park businesses, 13.5 percent provide services to these firms, and 12.4 percent buy supplies from organizations in the park.

IMPACT ON FIRM LOCATION AND
EMPLOYMENT GROWTH

Clearly, the Stanford Research Park has been a major factor in the rapid growth of the San Mateo–Santa Clara bicounty region. The park itself employs a considerable number of workers who might not otherwise be in the region. Moreover, park businesses have spun off other enterprises in the area. Out-of-park businesses are responsible for a significant amount of economic activity and employment either because they exist solely to serve or be served by park organizations or because they have significant trading relationships with them. This includes the creation of jobs in sectors other than high technology that have been established to satisfy the additional consumer demand stimulated by park and park-induced growth.

Our survey of park businesses reveals that roughly six out of the more than fifty businesses in the park would not be located in the region if the Stanford Research Park did not exist, and that two businesses would not exist at all. This translates into some 1,700 jobs in the region and 750 jobs anywhere. Our survey of out-of-park businesses indicates that roughly 350 firms, or 22,000 jobs, would not be in the two-county region if the Stanford Research Park did not exist.[42]

It is interesting to compare these estimates to those for the regions surrounding the University of Utah and Research Triangle parks. In the Stanford region, a relatively small proportion (about 6 percent) of in-park jobs would not exist without the park, compared to 21.5 percent of the jobs in the University of Utah Research Park and over 60 percent in the Research Triangle Park. This attests to the economic strength and attractiveness of the Palo Alto area and the importance of the university connection independent of the research park.

The proportion of the high-tech base made up of induced out-of-park businesses is higher in California (2.3 percent) than in North Carolina (1.2 percent) but lower than in Utah (3.4 percent). The higher ratios in the Palo Alto and Salt Lake City regions, compared to the Research Triangle, may be due to the higher incidence of spin-off activity in the former regions, although nonresponse bias from the survey cannot be ruled out.

These estimates of induced job creation do not include employment growth resulting from (1) induced spending on consumer goods and services by employees and their families in park and out-of-park businesses, and (2) "induced" park organizations' purchases of inputs from other local businesses. [43] Expenditures on consumer items and intermediate inputs that would not be made without the park create jobs that also might not otherwise exist in the region. We estimate that approximately 43,000 jobs have been generated in the San Mateo–Santa Clara region due to the household income multiplier. We further estimate that 8,100 jobs have been generated by the local purchases of all businesses in the region that would not be in the region if the park did not exist. The sum of all jobs that currently owe their existence to the presence of the Stanford Research Park is approximately 75,000. That represents 3.7 percent of all jobs in the two-county region. The proportion is smaller than that for the Raleigh-Durham area because the size of the economic region is considerably larger, but it is larger than that for the University of Utah region probably because of the latter's relatively young age.

Another way we attempted to determine the park's impact on economic growth was to ask representatives of park organizations, the park manager, and university officials about their own perceptions. Over half the park business respondents, as well as the park manager and all but one university official, believed that employment growth and business start-ups in the region would have been either somewhat or much lower without the park.

It is important to note that the regional growth stimulus from the Stanford Research Park, and from the other parks, is not over. Almost two-thirds of the businesses in the Stanford and Research Triangle parks expect that their organizations will continue to grow over the next five years, and almost 90 percent of the businesses in the younger University of Utah Research Park project further growth.

IMPACT ON INCOME GROWTH

Between 1950 and 1987, per capita personal income in Santa Clara and San Mateo counties grew from 135.5 percent to 147.7 percent of U.S. per capita income. The more rapid income growth in the region versus the nation can be attributed, in part, to the presence of the research park. We already have established that a measurable share of the region's employment growth can be traced to the park. Many of those jobs are in high-tech businesses that, on the average, employ a higher proportion of skilled and professional workers than the modal business found elsewhere in the United States. On the average, those high-tech businesses pay higher salaries.

To assess the impact of the park on income growth in the region, we compared income growth in the park's county (San Mateo) to income growth in control group counties over different time periods. In every decennial period since 1950, the per capita personal income growth in the Stanford area has been more rapid than that for the control group.

Respondents' perceptions about the effect of the park on wage and salary growth for professional and nonprofessional employees were somewhat mixed. Of the twenty-seven in-park business respondents, only eight felt that compensation would be lower without the park. The others believed that it would be about the same. One-half of the university officials surveyed felt that compensation would be lower, and one-half thought that it would be about the same.

IMPACT ON THE LOCAL LABOR MARKET

In 1950, before the Stanford Research Park existed, income inequality in Santa Clara County, as measured by the GINI coefficient, was virtually the same as it was nationwide.[44] During the first half of the 1960s, regional income inequality worsened in Santa Clara and San Mateo counties—in absolute terms and relative to the index for the country. (See Figures 7-7 and 7-8.)

Figure 7-7. *GINI Coefficients of Income Inequality in the Stanford Research Park Region*

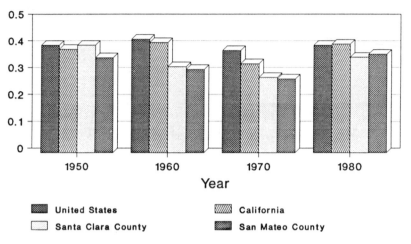

Sources: U.S. Bureau of the Census, *1950, 1960, 1970, and 1980 Census of Population.*

We do not have hard evidence to link the trends in inequality to the growth of the Stanford Research Park and related activity. In a study of Silicon Valley, Saxenian attributes the growing inequality since the mid-1960s to the bipolar nature of many of the high-tech industries in the region, so that highly paid R&D scientists and low-paid assembly workers are employed in close proximity to each other.[45] Therefore, to the extent that the park has contributed to the development of the high-tech economy in the region, it also has contributed to the inequality.

In face-to-face interviews with business representatives, park and university administrators, and local government officials, we were told repeatedly that an emerging problem for park organizations and other businesses in the area is the need for lower-paid workers to drive long distances to work because local housing prices are so inflated. If this is true, the income inequality referred to above understates the actual disparity, since lower-income workers are forced to live outside the metropolitan area.

Women make up 40 percent of the workforce within the park and 47.2 percent of the workforce outside the park. The former figure is below the national average, and the latter is above it. The smaller share of women workers in the park, compared to the region, state, and nation, can be accounted for by the occupational mix in the park, which is skewed more heavily toward managerial

Figure 7-8. *Ratio of County-to-U.S. Income Inequality in the Stanford Research Park Region*

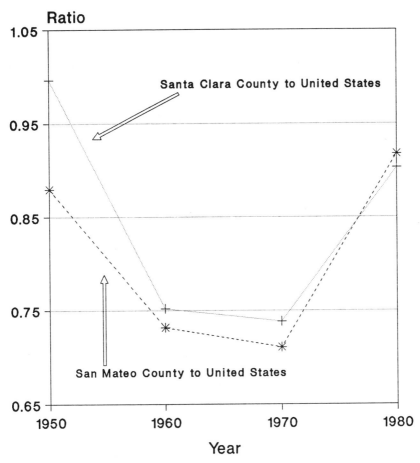

Inequality Measured by GINI Coefficients

and technical positions in which women are underrepresented. The relatively high representation of women outside the park is most likely a consequence of the occupational mix of the assembly operations in Silicon Valley.

As noted earlier, minorities comprise 20 percent of the park's workforce, whereas they represent 15.2 percent of the workforce of out-of-park businesses. Both proportions are above the national average. The higher share of minority workers in the park seems to be a result of the large Asian-American population in the Bay area and their overrepresentation in park occupations. The share of black and Hispanic workers in the park is less than for the state and the nation—for the same reason as women. The relatively high ratio of minorities outside the park is likely a reflection of the large Asian-American population as well as a high representation of blacks in the less-skilled occupations.[46]

As a way of judging the possible effect of the park and related industrial development on labor market outcomes for women and minorities, we traced the ratio of unemployment rates in the two-county region to those in California and the United States for these two groups. As shown in Table 7-7, compared to the state and the nation, the labor market position of women has improved steadily since the Stanford Research Park was established in 1951. The largest relative improvement came in the 1970s.

Minorities have not fared as well in the local labor market. Compared to the experience of minorities in California generally, the unemployment rate for nonwhites living in Santa Clara and San Mateo counties grew higher until 1970, after which time it improved relatively. But, as of 1980, the unemployment rate of minorities in the region was still worse than in 1950 compared to the rate for California as a whole. Compared to the nation generally, labor market outcomes for minorities in the region improved considerably over the 1950–80 period, but only because of a dramatic turnaround in the 1970s.[47]

We do not have enough information to draw causal inferences from these data on women and minorities. We can only speculate that the expansion of assembly-type jobs in the region, especially in the 1970s, created opportunities for the less-skilled members of these groups. If that is what has happened, there is some reason for pessimism in the future as the assembly-type jobs increasingly move offshore.

With regard to the park's likely impact on the occupational mix and local wage structure, the concentration in the region of high-tech businesses has resulted in a higher-than-average proportion of engineers, scientists, and skilled technicians—both within and outside the park. That development, plus the explosive growth of population in the region, has driven up wages and the cost of living across the board. The cost of housing is a good illustration. In 1950, the median price of a single-family home in Santa Clara County was $10,644; by 1980, it had risen to $110,500, or $23,156 in 1950 dollars.[48]

Stanford University and research park businesses together employ some 30,000 engineers, scientists, administrators, and managers. Out-of-park businesses in the region employ tens of thousands more. Given this demand, it is

Table 7-7. *Unemployment Rates for Women and Minorities in the Two-County Region versus California and the United States*

	Ratio of Unemployment Rates			
	1950	1960	1970	1980
Santa Clara and San Mateo Counties versus California:				
Women	0.96	0.91	0.88	0.67
Minorities	0.20	0.59	0.72	0.53
Santa Clara and San Mateo Counties versus the U.S.:				
Women	1.72	1.11	1.17	0.67
Minorities	0.61	0.68	1.08	0.37

not surprising that 65 percent of in-park respondents and 55 percent of out-of-park respondents cited the scarcity of professional workers as a problem. It also is not surprising that approximately half of these workers are imported from outside the area.

Regional businesses pay a premium for professional workers. In addition, there appears to be wage competition for skilled technicians, who also are in short supply. Most of these workers are hired locally—from two-year colleges or laterally from other businesses. The escalation in wage levels for professionals and skilled nonprofessionals was cited as a major problem by 40 percent of the out-of-park businesses. We have no evidence that this has resulted in industrial gentrification or in the failure of some businesses that cannot compete in this inflationary labor market, but the possibility of that occurring in the two-county area cannot be ruled out.

The demand for less-skilled workers is less intense, and there seems to be less upward wage pressure in these occupations. For example, only 14 percent of park businesses indicated that custodians were in short supply. There does appear to be some wage roll-out, however, since the ratio of wages paid to this group of workers to the U.S. average has grown over time. This wage inflation, however, seems to be less demand-driven than cost-of-living driven.

IMPACT ON THE REGION'S INNOVATIVE CAPACITY

It is difficult to separate the impact of the park from the impact of the university on the region's unusually high capacity to nurture and sustain innovative businesses and successful entrepreneurs. The park has had such an impact directly by attracting a number of businesses that otherwise would not have located in the region. In particular, the park has had a role to play in the region's well-deserved reputation for spawning start-up companies. Based on responses to the park business survey, park businesses are responsible for about 120 spin-off companies.[49]

Some of the strongest evidence of the park's effect on innovativeness comes from Niels Reimers, director of Stanford's Office of Technology Licensing. He provides many examples of how park businesses have helped to commercialize the inventions of university faculty or how faculty members have established a business in the park to commercialize inventions (recall Hewlett-Packard, Varian, and Watkins-Johnson). Increasingly, these businesses are being established or contracted with outside the park because park land is scarce and rents are high.[50]

In addition, the park has contributed indirectly to the innovative capacity of the region by helping to strengthen the university's research capacity and reputation. Another indirect contribution, but difficult to measure, is the park's effect on the worldwide reputation of the region as a center of innovation. This effect is manifest in the number of high-tech businesses that have come to the region to share in the substantial localization economies built by the park and the university.

IMPACT ON THE LOCAL BUSINESS CLIMATE
AND POLITICAL CULTURE

The research park, along with Stanford University, has contributed to a widespread perception of the region as a high-tech mecca. Many businesses have moved to the Palo Alto area, not because they have particular economic relationships with the university, park organizations, or even other high-tech businesses, but because they value the dynamic milieu or simply want to have a Stanford Research Park or a Silicon Valley address. This is revealed by the weight given to "business climate" in our survey responses, as well as by the fact that "the prestige of the park" is the second most beneficial aspect of park proximity, as judged by out-of-park businesses.

The Stanford Research Park also has had a significant effect on the political culture of the region. Many park professionals have been active in the local community, serving on appointed boards and commissions and in leadership positions in civic and business organizations. According to current city officials, this involvement was more frequent in the 1950s, 1960s, and early 1970s than it has been during the past fifteen years—for two reasons. First, park businesses increasingly have developed an international focus, and second, local politics has become somewhat polarized between business and development interests on one side and environmentalists and no-growth proponents on the other. Many executives of park businesses have not wanted to enter this fray.

Conclusion

If the Stanford Research Park had not been developed, the Palo Alto area would still have one of the world's largest concentration of high-tech industry. The evidence suggests, however, that the park has affected the quantity and rate of growth of the high-tech economy at the margin and has had major benefits for Stanford University.

The total number of jobs created in the region—both in businesses that probably would not exist without the park and in the consumer goods sector—is approximately 75,000, or 3.7 percent of the employment base in the two-county region. But the economic importance of the Stanford Research Park is larger than these induced employment figures indicate. The park also has contributed to a positive business climate, which has served to attract to the region not only businesses, but talented individuals as well. In addition, park businesses have paid more into town coffers in taxes than they have cost in service provision, thereby helping to finance public services for Palo Alto and Santa Clara County residents.

The park also has helped to shape the political culture of the region by introducing into the political arena a group of powerful corporate executives who have played an active role in civic and public affairs. Initially, many of these executives were from the Bay area, but in more recent years they have immigrated from other regions and countries.

When it was established, the research park's primary objectives were to generate revenue for the university and to bring industry and university researchers closer together. The park has accomplished both of those to a greater degree than could have been envisioned forty years ago. Moreover, the park has helped to create a world-class electronics sector in the region that has generated benefits for the entire nation. Because Stanford University still owns the land, it will continue to reap financial rewards from real estate appreciation, especially as older leases begin to expire in the twenty-first century.

The evidence also suggests that during the past two decades the park has contributed to widening disparities between occupational and income groups in the region. In the long run, these disparities may lessen as less-skilled jobs are exported offshore and as escalating housing prices force lower-income workers to move elsewhere. At the same time, some observers believe that R&D may be beginning to disperse from the region, so that job growth at the high end of the income scale will slacken.[51]

The success of the Stanford Research Park, both as a real estate venture and as an instrument of economic development, is due less to planning foresight than to fortuitous timing and the ability of university officials to be flexible in their development efforts. Stanford University happened to have a critical mass of inventive individuals who needed space near the university at the time that electronics and other high-technology industries were about to burgeon. And, because Stanford's park was among the first to be established, there was little competition by other leading research universities for these innovative businesses. In a benefit-cost sense, the development of the Stanford Research Park seems, on balance, to have been positive. This is largely because the land on which it was located had limited options for its use, and neither the university nor state and local government had to provide financial incentives (other than arguably favorable lease terms) or expensive amenities to induce businesses to locate in the park.

The central lesson from this case study is that the experience of the Stanford Research Park is not likely to be replicated elsewhere. Few universities can match Stanford's excellence in technology-related fields, and few, if any, have as large an endowment of land. Few regions have the entrepreneurial tradition of the Palo Alto area. However, some of the elements that have made the park successful and given it its worldwide reputation—including close ties between the university and resident businesses in general and flexible park policies in particular—could be emulated and, in fact, appear in other successful parks.

8 Knowledge for Profit

The University— Research Park Connection

The previous chapters have described many of the significant linkages between research parks and universities specifically in terms of universities' ownership and operation of research parks and the importance of university proximity for the success of research parks as regional growth stimuli. A related issue is how, and to what extent, research parks create benefits and opportunities for universities and their faculties. This is important for regional economic development because universities can play a major role in the development of a region's innovative capacity.

One theme in this discussion is that, increasingly, university administrators and faculty regard research park development as one of several strategies that will strengthen partnerships between universities and private industry. This recognition by university administrators that private businesses will pay to locate near their research facilities, faculty members, and graduates is the basis for this chapter title, "Knowledge for Profit."

Universities and Research Parks: The New Institutional Stance

As many as one hundred U.S. universities currently are affiliated with a research park in some way; other universities, colleges, and technical institutes are considering a research park strategy. In general, the willingness of universities to affiliate with a research park reflects a new attitude among administrators and many faculty members about the appropriate role of their institutions. The traditional view—still held by some academics, particularly in the humanities and the basic sciences—is that universities should engage in "pure research" and strictly intellectual activities.[1] The National Science Foundation, for instance, still requires its applicants for research grants to demonstrate the con-

Table 8-1. *University-Sponsored Technology Development Activities*

Activity		University Type*					%
		I	II	III	IV	Total	
Provide patenting/ licensing assistance	No	0	2	8	5	15	
	Yes	33	13	16	13	75	83.3
	NA	0	0	0	0	0	
Participate with industry in joint research	No	4	0	5	6	15	
	Yes	28	13	19	12	72	82.8
	NA	1	1	0	0	2	
Provide technical/ managerial assistance	No	9	6	13	15	43	
	Yes	24	8	11	3	46	51.7
	NA	0	0	0	0	0	
Construct/operate specialized labs or other facilities	No	22	6	15	10	53	
	Yes	11	7	9	8	35	39.8
	NA	0	1	0	0	1	
Sponsor research parks	No	15	8	16	16	55	
	Yes	18	6	8	2	34	38.2
	NA	0	0	0	0	0	
Construct/operate small business incubators	No	18	10	22	14	64	
	Yes	15	3	2	4	24	27.3
	NA	0	1	0	0	0	
Participate in small business assistance centers (SBACs)	No	26	10	20	15	71	
	Yes	3	1	4	3	11	13.4
	NA	4	3	0	0	7	
Make equity investments	No	25	12	23	18	78	
	Yes	7	1	1	0	9	11.5
	NA	1	1	0	0	2	

* Carnegie Foundation for Higher Education classifications. Types I and II are research universities; Type I universities have a greater amount of externally funded research and award more Ph.D.s. Type III are doctoral-granting universities, and Type IV are degree-granting engineering and medical institutions.

tribution of their proposed activity to "basic knowledge," as opposed to public policy or product commercialization. Increasingly, however, there has been a new institutional stance that recognizes the importance of R&D alliances and partnerships with industry, not only to help make industry more competitive internationally, but also to bring badly needed revenue into the university, to retain entrepreneurial faculty, and to contribute to the economic development of the region.[2]

The creation, operation, or other form of affiliation with a research park is but one of several possible strategies universities can adopt to bind themselves more closely with private industry. Table 8-1 summarizes responses from a sample survey of vice-presidents for business and finance (or an equivalent officer) in eighty-nine institutions of higher learning. The purpose of the survey was to ascertain the type and extent of university-industry activities that have potential benefits for technology development or regional economic development. Specifically, we asked whether the universities engaged in the following activities:

1. Provide patenting and/or licensing assistance to faculty inventors

2. Provide technical and/or managerial assistance to small technology-oriented businesses

3. Participate in Small Business Assistance Centers (SBACs)[3]

4. Use university endowment funds to purchase an equity position in small technology-oriented businesses

5. Construct/operate specialized laboratories or other facilities

6. Construct/operate small business incubators

7. Participate with private industry in joint research

8. Sponsor research parks

As shown in Table 8-1, 38.2 percent of the respondents, or thirty-five universities, sponsor research parks.[4] But many more universities engage in other activities that are technology development– or economic development–oriented. For instance, over 80 percent of the universities surveyed engage in joint research with industry, whether or not those businesses are in nearby research parks. A similar percentage of universities helps faculty to obtain licenses and patents for their inventions. The table indicates that those and other activities are more likely to be sponsored by Type I research universities than by other types of institutions.

Universities as Research Park Owners and Operators

Research parks can be considered a prototype of the university-industry connection. They are but one of eight types of technology development activities that universities sponsor (see Table 8-1). Research parks typically are larger in scope (for example, they require more capital investment and employment) than the other university-sponsored initiatives and therefore have the potential for more sizable consequences (positive and negative) for the university and other "stakeholders." Because research parks are highly visible, they take on a larger political/symbolic meaning both within and outside the university. We suspect that many of the insights gained from this study of universities' affiliation with research parks also apply to other nontraditional university economic and technology development activities.

Universities play a formal role in the ownership, administration, and/or operation of approximately 60 percent of all research parks in the United States (see Appendix A). About 25 percent of all parks are units of the university; another 35 percent are governed by universities in other ways, sometimes as joint ventures with private developers (in 21 percent of all parks).

As the results of our large sample survey and case studies indicate, universities associate with research parks for different reasons and in different ways. In general, managers of university-owned research parks attach considerable importance to technology development as an objective of research parks (see Table 3-1). The specific goals that they cite include (1) to enhance the university's technical training capability via collaborative research, (2) to increase technology transfer, (3) to encourage entrepreneurship (for example, the University of Utah Research Park), (4) to increase regional productivity through innovation, (5) to generate revenue via land sales and leases (the primary objective of the Stanford Research Park), (6) to enhance the quality and prestige of the university, and (7) to commercialize university-based research. University officials—especially in state-supported institutions—also cite economic development as a goal of their research parks (for example, the University of North Carolina campuses involved with the Research Triangle Park). The more specific economic development objectives often include diversification of the region's economic base, development and nurturing of new businesses, expansion of local employment opportunities, and provision of higher-paying jobs.

The organizational links between universities and research parks are as varied as the motivations universities have for associating with the parks. Some universities own and operate the parks as they do other land and building holdings. Other universities help establish the park but then turn over its management to a subsidiary unit. Still others simply send representatives to sit on the board of an independent nonprofit organization. Finally, many universities have no formal role in park ownership or operation but maintain "informal" ties to nearby parks primarily through the activities of individual faculty members or academic departments. For example, the private Biomedical Research and Innovation Center in Miami, Florida, identified Florida International University and the University of Florida as important institutions in their operation; the Geddes Center in Superior Township, Michigan, has ties to Eastern Michigan University; Interstate Research Park in Champaign, Illinois, considers the nearby University of Illinois to be important; St. Paul Energy Park has connections with the University of Minnesota; and so on. (Appendix A contains further examples.)

The type of university-research park relationship in a particular case depends on several factors, including the history of park land ownership, the internal governing structure of the university, the robustness of the local economy, and the influence of key individuals within and outside the university. Just as Stanford University developed its research park, in part, as a way to put idle land to use, more than forty other universities also created research parks as a means of using university lands. One critical determinant of park ownership and operation is the university's internal structure. Smaller, non-research universities are less likely to have the capacity to manage a research park effectively. Those institutions are more likely than large research universities to turn over operations to a third party. Another factor is location. On the average, parks owned and operated by private, for-profit businesses have been established in the fastest-growing regions of the United States, suggesting that private developers limit their park development activity to the "hottest" real estate markets.[5]

Universities as an Ingredient of Park Success

We theorized in Chapter 3 that regions, especially smaller ones, are more likely to have growth induced by research parks if a university (or other research-oriented institution) is present. This reflects the fact that universities contribute to the creation of "localization economies" for park businesses or the presence of resources that businesses of the same type (concentrated in research parks) need, including specialized laboratory facilities, faculty expertise for consulting, and a regular supply of graduates for entry-level professional jobs. These university-generated localization economies are especially important in smaller regions, which are not likely to have other sources for generating external economies. Businesses in or near university-related research parks also can benefit from the prestige of association: location in the Stanford, Princeton Forrestal, University of Utah, and other such parks is particularly useful for businesses in the research or "knowledge" business. Finally, we argue that of two regions with research parks, the one also having a university will grow faster because of the greater likelihood that spin-offs or start-up companies will be established by faculty and graduates who choose to remain in the area and find research parks to be appropriate location sites.

While the synergies between universities and research parks play an important role in the growth-creation process, it is difficult—empirically—to separate universities' growth-inducement effects that operate through research parks from the economic stimulus from more general university activities. In short, universities, not research parks, may be the "growth pole." There is now a sizable literature on the regional economic benefits of university research. Nelson points out that university research "enhances technological opportunities and the productivity of private R&D in a way that induces firms to spend more both in the industry in question and upstream." Nelson found that university research affected innovation in industries related to the biological sciences. Similarly, Bania, Eberts, and Fogarty argue that university research spending affects the incidence of local start-ups, and Jaffe hypothesizes that university research spending affects corporate patenting activity. Finally, Florax and Folmer provide statistical estimates of the relationship between universities and different types of investment in regions. They conclude that proximity to a university "might be an important location factor . . . for investment in buildings." [6]

Presented in Table 8-2 is a taxonomy of the potential economic impacts of universities. This taxonomy is intended to help delineate the range of potential impacts and suggests some of the conditions under which the impacts are most likely to occur. (We do not attempt to measure the magnitude of the impacts here.) Some of the impacts occur through park businesses, others occur through businesses that are not in the park but locate in the region because the park exists, and still others occur through businesses not related to the park at all.

The five kinds of potential growth impacts included in the table come from (1) university expenditures on payroll, purchases, and taxes or in-lieu-of-tax payments, which would stimulate regional demand for goods and services and, consequently, for labor and capital via a multiplier process; (2) the provision of knowledge and training to students (human capital investment) and expertise to local businesses, which would enhance labor and general business productivity, respectively; (3) technology transfer activities, such as manufacturing modernization, which should increase the productivity and, hence, the competitiveness

Table 8-2. *Potential Impacts of Universities*

Types of Impacts	Types of Universities*	Geographic Scope**	Comments
SPENDING MULTIPLIERS			
Payroll	All	LOC	Spending impacts depend on size and structure of the local economy.
Nonlabor purchases	All	LOC, NATL	Balance tax (or in–lieu) payments with the opportunity cost of using land for a park. Net increase in payments can finance more/better public services.
Taxes (or in–lieu payments)	All	LOC	
KNOWLEDGE TRANSFER			
Faculty to students	All	LOC, NATL, INTL	Enhances productivity; depends on extent graduates stay in region.
Researchers to clients	I, II, IV	LOC, NATL, INTL	The larger, more prestigious a university is, the more international its impacts are likely to be.
TECHNOLOGY TRANSFER			
Licensing/patenting	I, II, IV	NATL	Depends on faculty inventiveness and marketing.
Technology centers	I, II, IV	NATL	Mostly help large firms.
Joint ventures	I, II, IV	NATL, INTL	May be limited by university restrictions; done mostly with large firms.
ASSIST START–UPS			
Technical assistance	I, II, IV	LOC, NATL	Affects small firms; the larger, more prestigious a university is, the more its expertise will be sought by national or international businesses.
Equity investments	Larger	LOC, NATL	May be governed by university restrictions.
Spin–offs	I, II, IV	LOC	University may restrict faculty from spinning off or may provide incentives. Also depends on availability of venture capital and the business milieu.
ATTRACT EXISTING BUSINESS			
Workforce recruitment	All	LOC	Depends on specialized labor needs of businesses and prestige of university.
Localization economies	All	LOC	Depend on the size of firms and the prestige of the university.
Create favorable cultural milieu	All	LOC	Depends on the spatial distribution of cultural activities in the region.

* For a definition of the four types of universities, see Table 8–1.
** LOC = Local; NATL = National; INTL = International.

of existing businesses in the region; (4) direct investments in, and technical assistance to, small business start-ups and faculty entrepreneurism, which should increase the rate of enterprise formation and decrease small business mortality in the region; and (5) attraction of businesses to the region that seek access to trained labor, expertise and facilities, and a favorable intellectual-cultural milieu—all of which can be provided by a university.

We compiled three kinds of evidence about the effect of universities on employment growth. First, we compared the average annual employment growth rates of counties in the United States with three different kinds of universities to counties that do not have universities. Second, we included measures of "university presence" in an analysis of the determinants of research park *success*, defined as a positive difference in employment growth rates between research park and other counties. And third, we asked top managers of technology-oriented businesses to indicate the importance of universities in their location decision.

The first type of evidence is presented in Figure 8-1. Counties with degree-granting medical and engineering institutions appear to have grown more rapidly than the control group of counties. Counties containing Ph.D.–granting (but not research) universities appear to have grown more slowly than others, and counties with Type I research universities seem to have grown at around the same rate as the control group of counties. Admittedly, this analysis is crude because it does not control for other causes of county growth besides the presence of a university—including the presence of research parks.

The second type of evidence was presented in Chapter 4, where we discussed results from multivariate statistical analyses that consider several possible sources of employment growth at once. Ordinary least squares and logit regressions and a hazards/survival model all indicate that counties containing research parks owned and operated by universities are more likely to grow faster than counties with research parks not owned and operated by universities.

The third type of evidence—the perceptions of business managers, both in and near research parks, about universities—was reported in the case study chapters (Chapters 5–7). In every survey we conducted, these managers reported that proximity to a research university was an important reason for locating in the region.

We can draw the following conclusions from this evidence about the importance of universities:

Figure 8-1. *Mean Annual Employment Growth Rate, by County, 1977–1987*

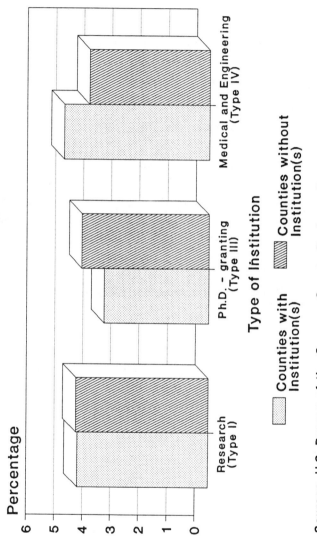

Sources: U.S. Bureau of the Census, *County Business Patterns, 1977 and 1987.*

1. The type of institution seems to matter in the growth-generation process. Counties containing the type of institutions most likely to produce commercializable products and business spin-offs—medical and engineering schools—have grown faster than the control group of counties, at least in terms of employment growth.[7]

2. The organizational relationship between the research park and the university seems to matter. Parks owned and operated by universities are more likely to generate growth than parks that are less formally affiliated with universities.

3. Businesses that have located near research parks have particularly strong attachments to the local university. This suggests that universities affect the *mix* of local businesses as much or more than they affect the *size* of the local business base. The effect of the university on employment and income, then, depends on the labor requirements and compensation patterns of the new self-selecting businesses relative to those that otherwise would have located in the region.

Research Parks as an Ingredient of University Success

The surveys we have conducted indicate a widespread perception that research parks benefit nearby and affiliated universities. More than 60 percent of the park manager respondents believed that their park has improved both the quality and the stature of the university. Similarly, administrators at Duke University, North Carolina State University, the University of North Carolina at Chapel Hill, the University of Utah, and Stanford University all indicated in their responses that the park near their campus has had positive effects on the quality and stature of their university as well as on its ability to generate sponsored research, attract good graduate students and place them after graduation, and attract and retain top faculty.

Stanford University officials gave a particularly strong endorsement of their research park, in part, reflecting the fact that the park has been a part of the university's land management operations for almost forty years. Administrators estimated that as many as five hundred graduate students receive money each year from park organizations in fellowships, part-time employment, and internships, and that fifteen to twenty scientists in park businesses serve as adjunct faculty, often with little or no compensation. University administrators also noted that park businesses help faculty inventors commercialize their products and contribute generously to the university both unrestricted gifts and facilities. This philanthropic dimension was also cited by administrators at the other case study universities.

We also asked some six hundred faculty members at the five universities affiliated with the Research Triangle, University of Utah, and Stanford parks to assess the benefits of the parks to them personally and to their respective institutions. The faculty sample was drawn randomly from selected departments with disciplinary foci related to the mission of park businesses, including economics and the policy sciences, engineering and design, mathematics and the physical sciences, health and the life sciences, agriculture, business, foreign languages, and computer science. Responses from the full sample are summarized in Table 8-3. The table also indicates whether the extent of the respondents' teaching or research activities, or their length of tenure at the university, is correlated with the responses.[8]

The most dramatic finding is that faculty members give their research parks an 86 percent approval rating. Over 40 percent of the faculty surveyed strongly approved of the university's relationship with the park. Approximately three-quarters of the respondents believed that the park enhances the quality and stature of the university in at least the following ways: by inducing the faculty member to accept the job offer (approximately one-half of the respondents indicated that they were influenced, at least somewhat, by the proximity of the park), by helping to recruit researchers to the university (70 percent of the respondents), by helping to attract top graduate students (47 percent of the respondents), and by helping to increase the interaction between scientists and researchers in the university and industry (70 percent of the respondents with opinions). The faculty members also indicated that they personally benefit from the parks. For example, almost 25 percent of the respondents had some involvement with park businesses, either in joint research or consulting or through the use of specialized facilities and equipment. And 15 percent of the eighty respondents who were involved with start-ups said that the park influenced their action.

The faculty survey results by academic discipline appear in Table 8-4. Based on these data, we can make the following observations:

Table 8-3. *Faculty Perceptions of Research Parks' Benefits for Universities*

Did the presence of the park influence you to accept the offer of a faculty position at this university? (N = 513)	Strongly 5.5% Moderately 12.1% Mildly 23.0% Not at all 59.0%

Comment: Positive responses are positively correlated with respondents' teaching load a negatively correlated with their longevity at the university.

Has the park met your expectations? (N = 508)	Yes, fully 18.1% Yes, partially 28.7% No, disappointed 9.8% No, very disappointed 1.8%

Comment: Most positive responses are from faculty who have been at the university less than ten years.

Do you have a current relationship with park organizations? (N = 592)	No 75.9% Yes 24.1%

Comment: 37.3% of respondents engaged in 5–9 hours of consulting per week responded "yes."

Has the park's existence helped your university recruit researchers? (N = 592)	Considerably 19.9% Some 49.2% Not at all 6.9% Don't know 24.0%

Comment: Positive responses are correlated with respondents' teaching load.

Has the park's existence helped you secure grants? (N = 581)	Considerably 2.6% Some 21.9% Not at all 57.8% Don't know 17.7%

Comment: The more active researchers rely the least on park businesses for assistance.

Has the park's presence helped you recruit graduate students? (N = 581)	Considerably 5.7% Some 41.8% Not at all 28.2% Don't know 24.3%

Table 8–3 (continued)

Has the research park served to increase the overall *quality of the university?* (N = 580)	Considerably 16.7%
	Some 59.3%
	Not at all 6.0%
	Don't know 17.9%

Has the research park served to increase the overall *stature of the university?* (N = 584)	Considerably 19.7%
	Some 54.6%
	Not at all 8.7%
	Don't know 17.0%

Comment: The more teaching and research done by respondents, the more positive their response.

Does the park's presence help graduates get jobs? (N = 578)	Considerably 14.0%
	Some 50.7%
	Not at all 10.4%
	Don't know 24.9%

Does the presence of the park increase the level *of interaction between the university and private* *businesses?* (N = 576)	Considerably 5.5%
	Some 35.6%
	Not at all 16.2%
	Don't know 42.7%

Do you approve of your university's relationship *with the research park?* (N = 586)	Strongly approve 41.1%
	Approve 44.7%
	Undecided 12.0%
	Disapprove 1.9%
	Strongly disapprove 0.3%

What percentage of your consulting time is with *park businesses?* (N = 468)	76 to 100 percent 10.0%
	26 to 75 percent 6.2%
	1 to 25 percent 21.6%
	Zero percent 62.2%

Did the research park influence you to start up *the business you did?* (N = 79)	Strong influence 5.1%
	Moderate influence 10.1%
	Mild influence 8.9%
	No influence 80.0%

Comment: 80 respondents, or 13.4% of the total, indicated they participated in a start-up.

Table 8-4. *Faculty Perceptions, by Discipline*

	Engineering and Computer Sciences		Math and Physical Sciences		Health and Life Sciences		Social Sciences, Languages, and Communications		Business and Architecture		Agriculture	
Have referrals by the university led to consulting opportunities?												
No	94	96.9%	64	91.4%	254	99.2%	63	100%	31	96.9%	49	94.2%
Yes	3	3.1%	6	8.6%	2	0.8%	0	0%	1	3.1%	3	5.8%
Total (N = 570)	97		70		256		63		32		52	
Percentage of total consulting time worked with organizations in the park:												
0%	56	62.9%	30	63.8%	120	57.4%	33	71.7%	12	48.0%	35	76.1%
1-25%	14	15.7%	10	21.3%	51	24.4%	10	21.7%	8	32.0%	7	15.2%
26-50%	5	5.6%	4	8.5%	11	5.3%	0	0%	2	8.0%	0	0%
51-75%	1	1.1%	0	0%	4	1.9%	0	0%	0	0%	2	4.3%
76-100%	13	14.6%	3	6.4%	23	11.0%	3	6.5%	3	12.0%	2	4.3%
Total (N = 462)	89		47		209		46		25		46	
Do you use any specialized facilities, equipment, expertise, or data in the park for research or teaching?												
No	69	71.9%	58	86.6%	207	80.5%	52	85.2%	27	84.4%	39	79.6%
Yes	27	28.1%	9	13.4%	50	19.5%	9	14.8%	5	15.6%	10	20.4%
Total (N = 562)	96		67		257		61		32		49	
Has the university generally been supportive of faculty working with organizations in the research park?												
Strongly	34	34.3%	24	35.8%	91	35.3%	13	22.0%	6	18.2%	16	30.8%
Somewhat	34	34.3%	20	29.9%	68	26.4%	11	18.6%	13	39.4%	17	32.7%
Neutral	25	25.3%	21	31.3%	81	31.4%	32	54.2%	13	39.4%	18	34.6%
Somewhat not	6	6.1%	1	1.5%	12	4.7%	1	1.7%	1	3.0%	1	1.9%
Strongly not	0	0%	1	1.5%	4	1.6%	1	1.7%	0	0%	0	0%
Don't know	0	0%	0	0%	2	0.8%	1	1.7%	0	0%	0	0%
Total (N = 568)	99		67		258		59		33		52	

Have you ever been a primary participant in a start-up company, either inside or outside of the park?

	Engineering and Computer Sciences		Math and Physical Sciences		Health and Life Sciences		Social Sciences, Languages, and Communications		Business and Architecture		Agriculture	
No	72	70.6%	67	97.1%	244	90.7%	58	92.1%	24	72.7%	44	84.6%
Yes	30	29.4%	2	2.9%	25	9.3%	5	7.9%	9	27.3%	8	15.4%
Total (N = 588)	102		69		269		63		33		52	

Has the existence of the research park nearby made it easier for you to commercialize your research?

	Engineering and Computer Sciences		Math and Physical Sciences		Health and Life Sciences		Social Sciences, Languages, and Communications		Business and Architecture		Agriculture	
Don't know	32	33.0%	24	35.8%	102	40.6%	27	45.0%	9	28.1%	8	17.0%
No effect	45	46.4%	37	55.2%	117	46.6%	29	48.3%	21	65.6%	30	63.8%
Some effect	15	15.5%	5	7.5%	27	10.8%	4	6.7%	2	6.3%	8	17.0%
Considerable effect	5	5.2%	1	1.5%	5	2.0%	0	0%	0	0%	1	2.1%
Total (N = 554)	97		67		251		60		32		47	

Do you feel that the university generally is supportive of faculty entrepreneurs?

	Engineering and Computer Sciences		Math and Physical Sciences		Health and Life Sciences		Social Sciences, Languages, and Communications		Business and Architecture		Agriculture	
Strongly agree	21	20.6%	10	14.5%	22	8.4%	6	9.8%	3	9.1%	3	5.9%
Somewhat agree	28	27.5%	21	30.4%	62	23.6%	12	19.7%	8	24.2%	7	13.7%
Neutral	35	34.3%	31	44.9%	113	43.0%	32	52.5%	9	27.3%	17	33.3%
Somewhat disagree	11	10.8%	3	4.3%	43	16.3%	8	13.1%	9	27.3%	14	27.5%
Strongly disagree	5	4.9%	4	5.8%	21	8.0%	2	3.3%	4	12.1%	10	19.6%
Don't know	2	2.0%	0	0%	2	0.8%	1	1.6%	0	0%	0	0%
Total (N = 579)	102		69		263		61		33		51	

1. Faculty members in the technical disciplines—engineering, computer science, math and the physical sciences—have found the park to be a greater draw in their employment decision than other faculty members.

2. Faculty members in the "softer" disciplines—social science, languages, and communications—are relatively more disappointed than other faculty members with what the park has provided, in part, because they tend to have less contact with park organizations, are helped less by park firms in securing grants, and place fewer graduates in park businesses.

3. Fifty-five of the seventy-nine start-ups indicated in the table (69.6 percent) were by faculty members in two groups—engineering/computer science and the health/life sciences. More than one-quarter of the faculty members in those two disciplinary groups who started new businesses indicated that the presence of the park had some effect on their decision.

There are some notable differences in faculty perceptions about the impact of research parks on the five universities. On the whole, a larger proportion of the faculty members at the three Research Triangle universities views the park as a source of benefit to their respective universities than do faculty members at Stanford and the University of Utah. The three Research Triangle universities rank first, second, and third in faculty perceptions that the park has improved the overall quality of their associated universities, helped recruit and retain faculty researchers, made it easier for faculty members to obtain research grants and contracts, and improved the ability to place graduates in jobs. Only in helping faculty members to commercialize their research and recruit high-quality graduate students do a higher proportion of Stanford and University of Utah faculty members feel more strongly than their RTP university counterparts about the benefits of their research parks (see Table 8-5). Faculty members at each of the three Research Triangle universities also give a higher approval rating to the relationship of their institutions with the research park than those at Stanford and the University of Utah, although those differences are not large.

The differential pattern by region is consistent with the relative magnitudes of the estimated economic impacts on the respective regions. In both sets of results, the Research Triangle Park represents a much larger resource compared to the size and scale of its region than is the University of Utah Research Park to the Salt Lake City metropolitan area or the Stanford Research Park to San Mateo and Santa Clara counties. Stanford faculty members have many other

Table 8-5. *Faculty Perceptions, by University/Region*

Proportion of faculty respondents indicating "considerable" or "some" impact of park on:	U.N.C.– Chapel Hill (%)	Duke Univ. (%)	NC State Univ. (%)	Stanford Univ. (%)	Univ. of Utah (%)
Improving overall quality of the university	75.4	71.8	82.9	61.3	71.4
Recruiting and retaining faculty	74.3	73.5	83.1	51.5	44.2
Helping faculty to obtain research grants or contracts	32.4	29.5	23.5	15.6	17.6
Recruiting top graduate students	57.7	40.3	59.8	45.5	18.0
Helping faculty commercialize their research	8.7	11.6	13.1	26.7	14.6
Increasing graduates' ability to get professional jobs in the region	73.7	62.0	71.9	61.3	42.8
Proportion of faculty approving relationship between the research park and their university	86.8	88.4	87.7	71.9	83.0

opportunities to collaborate with industry researchers and students have many other sources of jobs in addition to those in the research park. That is not the case in the Research Triangle area. In Salt Lake City, the University of Utah Research Park is small compared to the size of the university, and the businesses in the park are highly specialized in a relatively small number of technological fields.

Research Parks as a University Strategy

In the foregoing discussion, we have identified seven reasons why universities affiliate with research parks:

1. To help attract and retain entrepreneurial faculty.

2. To help attract good graduate students.

3. To increase collaborative research with private industry.

4. To facilitate technology transfer and the commercialization of faculty inventions.

5. To enhance the general quality and stature of the university.

6. To contribute to the economic development of the region.

7. To generate revenue through land sales and leases (for university-owned parks).

A variety of evidence summarized in this chapter suggests that these objectives are being met *in some instances*, in the perception of key actors, if not in fact. Most of this evidence comes from our case studies of three research parks. Because these parks were preselected for analysis because of the *a priori* perception that they were successful, we cannot generalize from our evidence about the benefits of all research parks to universities. In fact, along with the observation that some parks have succeeded as a real estate venture or as an economic development strategy is the recognition that other parks have failed. For example, several university-owned parks have few, if any, land sales or rentals and thus have been generating deficits for their university. In addition, spin-offs and other induced entrepreneurial activity seem to be limited to parks affiliated with certain types of universities. And, in Chapter 4, we reported that some parks have been more successful than others in generating jobs for their region.

We simply can confirm a seemingly self-evident maxim: that affiliation with a successful research park can be a successful strategy for universities. But even this conclusion must be qualified. First, it reveals a circularity that the probability of research park success, as we have defined it, depends (among other things) on affiliation with a university. If universities wait for parks to be successful before affiliating with them, they will have diminished the chances of that success occurring. Moreover, if the issue for university officials is whether or not to *start* a park, there will be no record of success or failure to guide them.

The second caveat is that we have focused on gross, rather than net, benefits. Even "successful" parks in terms of job creation, technology transfer, increased opportunities for collaboration, and so on may be so expensive to own and operate that they generate net losses for the university. Similarly, parks that generate a positive cash flow from real estate operations must be evaluated against the return from alternative uses of park land or nonland university resources. In short, research parks may not be the "highest and best use" for university land or the most productive use of university resources.

Universities that seek to maximize the net benefits from their affiliation with a research park should attempt to externalize as much of the cost of operating that park as possible. For instance, parks that are owned and operated by private corporations, nonprofit foundations, and state and local governments, but that are proximate to the university, would still allow universities to achieve all except the last two objectives listed above yet with far less financial obligation and risk.

University planners must also understand that research parks require, at a minimum, "deep pockets" and patience. Even so, new research parks may not succeed because they are entering the game late. Many of the other university-based economic and technology development initiatives listed in Table 8-1 are less costly, can yield benefits more quickly, and do not require universities to compete for a dwindling number of R&D branch plants. We are not suggesting that those strategies are always superior to the creation of a research park but, rather, that universities may be able to achieve some of the same objectives in a less costly and risky way.

9 Conclusion

In the preceding chapters we have attempted to assess the potential of research parks as a regional development tool. We have been motivated in this task by the recognition that neither sound theoretical nor empirical bases existed to justify the burgeoning popularity of research parks in recent years, both in the United States and abroad. Specifically, the hasty adoption of a research park strategy by state and local governments, universities, and private developers has raised the following questions:

1. Can the nation support the recent rapid growth in the research park population?

2. Can locations that differ—for example, in size, economic base, and existing institutions—all support research parks, or are research parks sustainable only in particular types of locations in the United States?

3. Assuming that research parks are viable, at least as real estate ventures, are the economic benefits they generate for their regions likely to exceed the public costs of the parks' development or opportunity costs?

4. Are the economic benefits that research parks generate likely to be shared by all segments of the region's population?

The answers to these questions can help us evaluate how research parks compare to alternative means of fostering regional economic development.

In this chapter, we draw together the insights and evidence from our literature/theory review (Chapter 2), cross-sectional analysis (Chapter 4), three case studies (Chapters 5–7), and knowledge-for-profit analysis (Chapter 8) to answer each of the questions listed above. Along the way, we examine the "critical success factors" that potential developers of research parks should consider. We then discuss a research park strategy within the larger context of regional economic development and other approaches to high-tech development and employment generation.

Can the United States Support Additional Growth in the Research Park Population?

Eighty-four of the 116 research parks that existed in 1989 were established after 1981. Most of these new parks, and many of the earlier ones, still are underbuilt (see Table 4-2), so their owners continue to pursue prospective occupants.

Research parks now are found in forty-four states. They contain, in the aggregate, more than 1,500 businesses, most of which perform high-tech research and development, and about 150,000 workers, many of whom are highly skilled. States with parks, and most remaining states without them, have indicated a desire to invest in even more research parks.

The efforts of existing parks to grow, and of state and local governments, universities, and private investors to create new research parks, come at a time when the growth of R&D activity in the United States is slowing down, as a trend and cyclically.[1] At least two scenarios are possible. If existing R&D establishments outside research parks do not relocate inside parks, the excess demand for R&D facilities by new and expanding parks will result in a continued high rate of park *failures*, as defined in Chapter 3. On the other hand, new and expanding parks could induce existing R&D facilities to relocate from nonpark locations. Some of these facilities might "hive off" from parent corporations, continuing the trend of spatial separation of corporate functions that has been occurring for some time.[2] Under this second scenario, the incidence of park failures would be lower than under the former scenario, but the public costs of relocation incentives would be higher. We then would have to ask whether the efficiencies gained from having agglomerations of R&D businesses within research parks, rather than in scattered locations, outweighed those additional public costs.

Are All Regions Suitable for Research Parks?

The literature we have reviewed and our empirical evidence both suggest that regions differ widely in their suitability for research park growth. In general, regions are most likely to host successful research parks if they have (1) an existing base of R&D and high-tech activity, (2) one or several research universities, medical schools, and/or engineering institutes, (3) good air service, (4) a well-developed network of infrastructure and business services, and (5) foresightful and effective political, academic, and business leaders. It is important to note, however, that *all* these elements need not be present. Some, like airports and infrastructure, can be affected by policy and may, in fact, develop as a result of research park growth.

Similarly, as a general rule, medium- and large-sized metropolitan areas seem to be the best locations for research parks because they contain large and diverse pools of labor and other external economies that both park directors and park businesses perceive to be important. Nevertheless, our evidence indicates that smaller areas with research universities, medical and engineering schools, and/or large federal laboratories also can support successful research parks.

The presence of these locational factors does not guarantee the success of a research park, however. As we have seen, vintage also matters. In the case of research park development, the early bird may, indeed, have caught the worm. Because the supply of new R&D facilities is growing at a slower rate than in the past, it is increasingly difficult for new parks to attract a critical mass of R&D activity. Footloose R&D facilities tend to prefer locations that already have been established as high-tech meccas.

In addition, park policies and procedures can affect the chances of park success. Parks that are owned and operated by universities, as opposed to just having a loose affiliation, seem to have stimulated more regional economic activity than other parks. (In Chapter 4, we referred to this as *linkage*.) And parks that provide amenities to park businesses seem to have become more successful than the others, all else being equal.

In sum, we conclude that the locational characteristics generally believed to make regions fertile seedbeds for research parks are neither necessary nor sufficient to ensure park success. The three successful parks in our case studies—the Research Triangle Park, the University of Utah Research Park, and the Stanford Research Park—all entered the game relatively early and were established in sizable regions near major research universities. Many other parks with similar characteristics that opened during the same period did not succeed, however. And several parks that appeared later in smaller regions have done well, at least by the measures of success employed in this study. Here, the lessons for planners and policymakers are threefold: (1) it will be increasingly difficult for *any* new research park to succeed, (2) it is relatively more difficult for a park to succeed if it is not closely linked to a university and is not located in a sizable region, and (3) certain barriers to park success may be overcome by good leadership, good luck, and good planning.

Do the Benefits of Research Park Development Exceed the Costs?

Using the framework of benefit-cost analysis, we would judge the desirability of a research park strategy by assessing its *net* economic benefits or, simply, the value of the induced direct and indirect economic benefits minus the associated actual and opportunity costs. As discussed previously, the induced benefits of park development include new wages paid, new household consumption, new purchases by induced businesses, and new tax payments made. The associated costs of this induced activity include front-end expenditures on land and infrastructure, as well as business recruitment, marketing, and operating and maintenance outlays. When money is not actually exchanged—for example, when land that is already owned is committed for use as a research park—there is still an associated opportunity cost if the land would earn a higher return in an alternative use.

The net economic benefits for our three case study research parks appear to be positive, but those cases are not representative of the full population of parks. We have demonstrated that the gross benefits are sizable, particularly for RTP and Stanford, because those parks are relatively large and old. At the same time, the associated costs (actual and imputed) for the three parks have been relatively low.

In terms of benefits, the Research Triangle Park has been responsible, directly or indirectly, for some 52,000 new jobs, 12.1 percent of the metropolitan area's total nongovernment employment in 1988, and three-quarters of the high-tech employment in the region (excluding the universities). The University of Utah Research Park has generated some 4,350 new jobs in the region that otherwise would not be there, or almost 2 percent of the metropolitan area total in 1988. Finally, the Stanford Research Park has created some 75,000 new jobs in San Mateo and Santa Clara counties, or 3.7 percent of that region's total employment. In the case of the Research Triangle Park, the regional payroll has been increased by almost $1 billion per year. The Salt Lake City regional payroll has been increased by approximately $70.5 million annually, and the Stanford area payroll is approximately $1.8 billion higher per year than it would be if the park had not been established.

In addition to these employment and income benefits, our case study research parks have helped to enhance the research capacities of their affiliated universities and to increase the rate of technology development, transfer, and diffusion. These benefits can be measured in terms of job and income creation, but not necessarily in the short or intermediate term, and they are not necessarily confined to the park's region. Moreover, each of our case study parks has had considerable symbolic importance by contributing to their respective region's reputation as a dynamic, high-technology center.

The benefits of research parks for universities have been documented in Chapter 8. There we have shown that administrators and faculty (especially in engineering and the "hard" sciences) in universities near successful research parks give those parks high approval ratings. The benefits that were cited include improvements in the universities' quality and stature, more and better jobs for graduates, easier access to research grants, and enhancement of the universities' appeal to researchers, new faculty members, and top graduate students.

On the cost side of the ledger, we must consider such outlays as land purchases, infrastructure development, and recruiting, marketing, operating, and maintenance expenditures. Direct outlays for land purchases and improvements vary from park to park. Those costs have been unusually low for the Stanford and Utah research parks since land purchases were not required. In Stanford's case, the university already owned the land, and in Utah's case, the federal government conveyed the land to the state, which in turn gave it to the university. RTP developers had to buy the acreage privately, but in the late 1950s land prices in the Research Triangle were still low by national standards. In each of the three cases, state and local governments assumed responsibility for essential infrastructure, thereby lessening the front-end costs for the developers. While that assumption by the government of responsibility for some infrastructure is fairly common, it is not automatic.

The opportunity cost of land for the three parks has been low as well. It is conceivable that RTP and Utah park lands could have been used for more intensive purposes (for example, manufacturing). We can only speculate whether those alternative uses would have grown more or less quickly than what we have actually observed. None of the alternative uses is likely to have generated more income per employee, however. In Stanford's case, there were few allowable alternatives for the land under the terms of the Stanford bequeathal.[3]

The front-end costs associated with the development of most other research parks either have been higher than for our case study parks or have taken longer to be paid off with proceeds from new business activities. This may account for many of the park failures described in Chapter 3. We noted that park development typically requires deep pockets and considerable patience. Many park developers simply do not have the luxury of either requirement.

One of the few generalizations that we can make about the net benefits of research parks is that they are far from certain. At best (as in our case studies), the net payoff from the investment occurs but it is slow to be realized. At worst, the park simply fails to achieve its promise and objectives. The probability of the latter outcome materializing seems to be growing as the supply of research parks has increased, especially relative to the supply of R&D businesses in the United States. Parks that pass through the incubation and consolidation stages into the maturation stage of development have the potential of generating considerable net economic benefits for their region. Yet, in the end, the performance of each park must be assessed individually.

Are the Benefits of Research Park Development Distributed Equally?

The last major question we have attempted to address in this study is whether, and to what extent, the economic benefits generated by research parks are shared among the segments of a region's population. In the case study regions, where we focused on the impact of research parks on regional income inequality, the local wage structure and types of jobs created, and opportunities for women and minorities, our findings regarding regional income inequality were mixed. In all three regions, as in the United States as a whole, income inequality worsened between 1970 and 1980. We found no evidence in the RTP and Utah cases to attribute any of this widening income distribution to park development. Circumstantial evidence in the Stanford case did lead us to suggest that park development contributed to income inequality in the region. The major difference between the Stanford Research Park and the other two parks that might account for this inequality is that the Santa Clara–San Mateo region also contains a concentration of high-tech–related manufacturing (Silicon Valley) that has grown up around the research park, whereas a smaller proportion of production jobs have been generated by the Research Triangle and University of Utah parks.

The income disparities are readily apparent in the Stanford region. High housing prices and taxes have driven many lower- and middle-income households to peripheral locations. People in some of those households commute long distances to their jobs in the research park—at considerable cost and inconvenience. Others are employed in lower-paying businesses outside the region. This last group of workers, at least, has been hurt economically, rather than helped, by park development.

It is possible that the spatial sorting out observed in California is occurring in Utah and North Carolina as well. That would not show up in our income inequality indices if nonprofessional workers were fleeing high costs of living near the parks by relocating in adjacent counties outside the defined region. There also may be a time lag related to the more recent vintages of RTP and the University of Utah Research Park and their somewhat slower rate of development to the maturation stage. If that is the case, then the *1990 Census of Population* data may indicate the beginning of the negative effects that are already evident in northern California.

Nonprofessional workers also would have an incentive to relocate away from the park if their wages did not rise proportionately with the cost of living. Our data indicate that wages for all workers have increased as a result of park development, but not by the same degree. The largest increases have been for professional and skilled workers, who have been in relatively short supply in each of the three case study regions. Parks have responded to the scarcity of professional workers by importing them from outside the region. The parks hire most of their skilled and semiskilled workers from area schools and, laterally, from area businesses.

Park development has increased opportunities for women and minorities in all three cases simply by broadening the supply of jobs. However, in each case, opportunities for white men have expanded more. In general, then, women and minorities are in a better absolute position but a worse relative position as a result of park development.

Our conclusion that the economic benefits from research park development have not been shared equally by all residents of the regions needs to be tempered with two facts. First, redistribution generally has not been a goal of the research park strategies developed by state and local governments or universities. And second, few regional alternative economic development strategies or programs rate high in terms of affecting income redistribution.

Research Parks and Alternative Economic Development Strategies

We have stated that research parks can be a costly economic development strategy in several respects. First, the expected failure rate is rather high (50 percent of start-ups fail, and 50 percent of surviving parks change their focus).[4] Therefore, the fact that the successful parks surveyed seem to have positive benefit-cost ratios needs to be tempered with the reality that for every two other announced parks, resources spent on park development are essentially wasted, and in one case per successful park, resources are spent on a park that ultimately will not meet the strategy's objectives.

Second, from the point of view of state or local governments in regions with small populations and without research universities or large government-sponsored laboratories, the probability of successful park development may be even lower than 0.25, especially for parks just starting up.

Third, even parks that achieve real estate viability may not generate the economic development outcomes that are sought. We have shown that counties hosting research parks have had a slower average rate of manufacturing job growth than similar counties without research parks. Of course, research park counties may still grow faster than control group counties in the longer run, as increased technology development and technology transfer and enhanced university research capacity begin to create new jobs, but those long-term effects may not be realized locally or in time to matter to elected politicians.

We also have shown that the jobs research parks create are suited disproportionately to better-educated workers. Because research parks are land-intensive and serve as locational magnets for other businesses and residences (if they are successful), they tend to raise land rents in the region. That negatively affects the region's less-skilled workers whose incomes have not been raised as a consequence of the park and related development. Park-related growth benefits white males relatively more than women and minorities, since white men tend to occupy the types of jobs that park businesses offer.

Finally, we have shown that many research parks are unlikely to be appropriate for new start-up businesses, because minimum lot size requirements and high unit land prices make the cost of entry into parks high. Not surprisingly, a large proportion of park tenants are branches of major national or multinational firms.

When deciding to create the Research Triangle Park, state economic development leaders hoped that high-technology production facilities would locate in adjacent but less costly regions of the state to take advantage of proximity to the R&D facilities in the park. That also has been an objective of other more recently established parks, particularly in states undergoing economic restructuring. Although unreconstructed growth center theory would support the premise that such a pattern of development should occur, we have found little evidence of it. Possible reasons range from overemphasizing the economic benefits of close spatial proximity of R&D and manufacturing facilities to other barriers to economic development in peripheral areas that first must be overcome, such as low skill levels of workers and inadequate education and training programs. We need to remember that, when successful, research parks will attract R&D activities to a region more than other business functions. Many of those who will get the new jobs may have to be recruited from outside the immediate region. Adjacent regions, rather than being the recipients of new manufacturing activities, ironically may suffer their own brain drains. In any event, the likely lack of significant spatial spillovers from research parks should represent a sobering consideration to economic development officials who hope that major economic benefits will extend to peripheral areas and to manufacturing workers.

In light of these drawbacks, state and local governments interested in employment creation for workers along a wider range of skill levels should consider alternatives to research park development. To enhance technology development and transfer between universities and private businesses, states could fund university-based R&D programs and centers. Such use of state funds would avoid the need for investments in expensive infrastructure. State and local governments that seek to create jobs for less-skilled workers might channel funds into job training and education programs, which also are less capital-intensive than park development. They also could target aid (in the form of subsidized or guaranteed loans, equity investments, technical assistance, or favorable tax treatment) to existing or start-up businesses that are likely to employ the types of workers most in need of jobs. Further, state and local governments could help new start-up businesses that would benefit from the kinds of external economies found in research parks by sponsoring business incubators and low-cost satellite parks. Alternatively, park owners could provide more multitenant buildings in their parks.

Our purpose here is not to recommend any particular alternative strategy but only to note that alternative programs exist. Just as policymakers should conduct a careful analysis before committing to a research park strategy, they also should make a painstaking study of individual alternatives to determine the appropriate program mix.[5] This kind of benefit-cost analysis unfortunately has not been done often. However, as the pace of regional economic change continues to accelerate, and as state and local governments continue to assume more responsibility for economic development policy-making, sophisticated analysis will become increasingly vital.

Unanswered Questions and Policy Lessons

We began this research in 1987 with what we considered to be an ambitious research agenda: to analyze the relationship between the presence of research parks and regional economic development, measured in terms of employment and income generation. As often occurs in the course of conducting research on interesting policy topics, many additional questions were raised that had been either unanticipated or underemphasized at the outset.

One set of new questions concerns the effect of research parks on measures of economic development other than jobs and wages. These include the incidence of industrial gentrification as a result of research park development, park-induced changes in land uses in adjacent areas, and the impact of park development on traffic patterns and environmental quality.

Another set of questions involves the role and effectiveness of universities in stimulating regional economic development. While conducting the empirical analysis, we suspected that in many cases the university itself, rather than the affiliated research park, was the more important engine of economic stimulation. We also were unsure whether university-sponsored economic development programs, including research parks as well as technology transfer programs, university-industry research projects, and small business assistance, or the traditional human capital-building function of the university were more effective in stimulating the regional economy.

The overall policy lesson we have drawn from this analysis is that in many regions research parks by themselves will not be a wise investment. The success rate among all announced parks is relatively low. And to the extent that vintage matters, it is too late for regions contemplating parks to get in on the ground floor. Research parks will be most successful in helping to stimulate economic development in regions that already are richly endowed with the resources that attract highly educated scientists and engineers. This is not to say that regions with less rich endowments cannot have a high-technology future, but more basic and long-term investments in improving public and higher education, environmental quality, and residential opportunities will be needed first. If a decision to create a research park is made, government leaders should be prepared to invest liberally, and all other stakeholders should be prepared to wait a number of years before the investment is returned.

Appendixes

Appendix A. Research Park Directory

Name of Park	Location	Year Opened	Nearby/Affiliated Universities*
Ada Research Park	Ada, OK	1960	East Central University
Advanced Technology Development Center	Atlanta, GA	1980	Georgia Institute of Technology
Ann Arbor Technology Park	Ann Arbor, MI	1983	University of Michigan
Arizona State University Research Park	Tempe, AZ	1984	Arizona State University
Athens Innovation Center and Research Park	Athens, GA	1987	University of Georgia
Baird Research Park Foundation Incubator	Buffalo, NY	1988	SUNY at Buffalo
Biomedical Research and Innovation Center	Miami, FL	1985	Florida International University, University of Florida
Bridgeport Innovation Center	Bridgeport, CT	1985	University of Bridgeport
Carolina Research Park	Columbia, SC	1983	University of South Carolina
Center for Business Innovation	Kansas City, MO	1985	University of Missouri – Kansas City
Central Florida Research Park	Orlando, FL	1979	University of Central Florida
Center for Advanced Technology	Ft. Collins, CO	1985	Colorado State University
Charleston Research Park	Charleston, SC	1984	Medical University of South Carolina, The Citadel, College of Charleston
Chicago Technology Park	Chicago, IL	1984	University of Illinois at Chicago
Clemson Research Park	Clemson, SC	1984	Clemson University
Colorado University East Campus Research Park	Boulder, CO	1987	University of Colorado
Connecticut Technology Park	Storrs, CT	1982	University of Connecticut
Cornell Research Park	Ithaca, NY	1951	Cornell University
Cummings Research Park	Huntsville, AL	1962	University of Alabama at Huntsville
Dandini Research Park	Reno, NV	1986	University of Nevada – Reno
Engineering Research Center	Fayetteville, AR	1980	University of Arkansas
First Coast Technology Park	Jacksonville, FL	1986	University of North Florida
Florida Atlantic Research and Development Park	Boca Raton, FL	1985	Florida Atlantic University
Geddes Center (Huron Center)	Superior Twp, MI	1981	Eastern Michigan State University
Great Valley Corporate Center	Malvern, PA	1974	Penn State University
Hannah Technology and Research Center	East Lansing, MI	1985	Michigan State University
Hawaii Ocean Science and Technology Park	Honolulu, HI	1985	University of Hawaii
Hidden River Corporate Park	Tampa, FL	1985	University of South Florida

Appendix A (continued)

NAME OF PARK	LOCATION	YEAR OPENED	NEARBY/AFFILIATED UNIVERSITIES[*]
Horn Rapids Business Park	Richland, WA	1982	Washington State University Extension
I.A.M.S. Research Park	Cincinnati, OH	1982	University of Cincinnati
Idaho State University Research Park	Pocatello, ID	1986	Idaho State University
Innovation Center and Research Park	Athens, OH	1978	Ohio University
Innovation Park	Tallahassee, FL	1978	Florida State University, Florida A&M University
Interstate Business Park	Tampa, FL	1983	University of Fort Lauderdale
Interstate Research Park	Champaign, IL	1963	University of Illinois
Iowa State University Research Park	Ames, IA	1988	Iowa State University
Johns Hopkins University Research Park	Baltimore, MD	1984	Johns Hopkins University
Kansas State University TechniPark	Manhatten, KS	1983	Kansas State University
Langley Research and Development Park	Newport News, VA	1966	George Washington University, Golden Gate University, Embry Riddle
Maine Technology Park	Orono, ME	1987	University of Maine at Orono
Maryland Science and Technology Center	Adelphi, MD	1982	University of Maryland
Massachusetts Biotechnology Research Park	Worcester, MA	1984	Worcester Polytechnic Institute, University of Massachusetts Medical School
Memphis Biomedical Zone Development Corp.	Memphis, TN	1988	University of Tennessee
Metropolitan Center for High Technology	Detroit, MI	1983	Wayne State University
Metrotech	Brooklyn, NY	1988	Brooklyn Polytechnic University
Miami Valley Research Park	Kettering, OH	1981	University of Dayton, Wright State University
Milwaukee County Research Park Corp.	Milwaukee, WI	1987	University of Wisconsin – Milwaukee
Minnesota Technology Corridor	Minneapolis, MN	1982	University of Minnesota
Mississippi Research and Technology Park	Starkville, MS	1984	Mississippi State University
Missouri Research Park	St. Louis, MO	1985	University of Missouri – St. Louis
Montana State University Technology Park	Bozeman, MT	1987	Montana State University
Morgantown Industrial and Research Park	Morgantown, WV	1973	West Virginia University
New Haven Science Park	New Haven, CT	1981	Yale University
New Mexico Institute of Mining and Technology Research Park	Socorro, NM	1985	New Mexico Institute of Mining and Technology
New Mexico State University Research Park	Las Cruces, NM	1986	New Mexico State University
Northwestern University/Evanston Research Park	Evanston, IL	1985	Northwestern University
Northern Kentucky University Research/Technology Park	Highland Heights, KY	1979	Northern Kentucky University

Appendix A (continued)

Name of Park	Location	Year Opened	Nearby/Affiliated Universities[*]
Oakland Technology Park	Southfield, MI	1983	Oakland University
Ohio State University Research Park	Columbus, OH	1984	Ohio State University
OREAD West Corporate and Research Park	Lawrence, KS	1986	University of Kansas
Oregon Graduate Center Science Park	Beaverton, OR	1982	Oregon Graduate Center
Oxmoor Valley High Technology Park	Birmingham, AL	1981	University of Alabama – Birmingham
Pennington Biomedical Research Park	Baton Rouge, LA	1986	Louisiana State University
Pittsburgh Technology Center Corp.	Pittsburgh, PA	1985	Carnegie–Mellon University, University of Pittsburgh
Princeton Forrestal Center	Princeton, NJ	1975	Princeton University
Progress Center	Alachua, FL	1985	University of Florida
Pueblo Memorial Airport Industrial Park	Pueblo, CO	1965	University of South Colorado
Purdue Industrial Research Park	West Lafayette, IN	1961	Purdue University
Rensselaer Technology Park	Troy, NY	1982	Rensselaer Polytechnic Institute
Research Forest	The Woodlands, TX	1984	University of Texas – Austin, Texas A&M University, Rice University, University of Houston – University Park
Research Park	Columbia, MO	1970	University of Missouri – Columbia
Research Triangle Park	Research Triangle Park, NC	1959	Duke University, University of North Carolina – Chapel Hill, North Carolina State University
Richland Industrial Park	Richland, WA	1962	University of Washington, Washington State University, Eastern Washington University
River Bend	Ft. Worth, TX	1976	University of Texas at Arlington
Riverfront Research Park	Eugene, OR	1985	University of Oregon
Rivers Corporate Park	Columbia, MD	1986	Johns Hopkins University, University of Maryland
Rochester Institute of Technology R&D Park	Rochester, NY	1986	Rochester Institute of Technology, University of Rochester
Rochester Science Park	Rochester, NY	1987	Rochester Institute of Technology
Roswell Test Facility	Roswell, NM	1983	New Mexico State University
St. Paul Energy Park	St. Paul, MN	1982	University of Minnesota
Shady Grove Life Sciences Center	Rockville, MD	1976	University of Maryland
Southgate University Park	Stony Brook, NY	1987	SUNY at Stony Brook
Stanford Ranch/Atherton Technology Center	Rocklin, CA	1986	California State University – Sacramento

Appendix A (continued)

Name of Park	Location	Year Opened	Nearby/Affiliated Universities*
Stanford Research Park	Palo Alto, CA	1951	Stanford University
Sunset Research Park	Corvallis, OR	1983	Oregon State University
Swearingen Research Park	Norman, OK	1950	University of Oklahoma
Synergy Research Park	Richardson, TX	1982	University of Texas – Dallas
Tampa Technology Park	Tampa, FL	1984	University of South Florida
Technology Innovation Center	Iowa City, IA	1984	University of Iowa
Tennessee Technology Corridor	Knoxville, TN	1982	University of Tennessee
Texas A&M Research Park	College Station, TX	1985	Texas A&M University
Texas Research Park	San Antonio, TX	1984	University of Texas – San Antonio
Thomas More Center	Crestview Hills, KY	1984	Thomas More College
Toftrees Technology Park	State College, PA	1985	Penn State University
TRI Park	Appleton, WI	1986	Fox Valley Technical Institute
University of California – Irvine Park	Irvine, CA	1983	University of California – Irvine
University Center	Albuquerque, NM	1988	University of New Mexico
University Center R&D Park	Tampa, FL	1982	University of South Florida
University City Science Center	Philadelphia, PA	1963	University of Pennsylvania, Drexel Institute of Technology, Temple University
University Corporate and Technology Park	New Orleans, LA	1988	University of New Orleans, Gulf State Research Institute
University of Delaware Research Park at Lewes	Lewes, DE	1985	University of Delaware – Lewes
University of New Hampshire Technology Park	Durham, NH	1983	University of New Hampshire
University Park	Edwardsville, IL	1988	Southern Illinois University
University Park	Mt. Pleasant, MI	1982	Central Michigan University
University Park at Kansas City	Kansas City, MO	1986	University of Missouri – Kansas City
University Park at MIT	Cambridge, MA	1982	Massachusetts Institute of Technology
University Research Park	Northridge, CA	1986	California State University – Northridge
University Research Park	Charlotte, NC	1968	University of North Carolina – Charlotte
University Research Park	Madison, WI	1984	University of Wisconsin – Madison
University of Tennessee Space Institute Research Park	Tullahoma, TN	1978	University of Tennessee
University of Utah Research Park	Salt Lake City, UT	1970	University of Utah

Appendix A (continued)

NAME OF PARK	LOCATION	YEAR OPENED	NEARBY/AFFILIATED UNIVERSITIES[*]
Utah State University Research and Technology Park	North Logan, UT	1985	Utah State University
Virginia Tech Corporate Research Center	Blacksburg, VA	1985	Virginia Polytechnic and State University
Washington State University Research and Technology Park	Pullman, WA	1981	Washington State University
Westgate/Westpark	McLean, VA	1982	
Youngstown Commerce Park	Youngstown, OH	1983	Youngstown State University

[*] Not all of these universities and institutes are formally affiliated with research parks, but have been identified by park managers as being important to their parks. The universities' and institutes' names appear in the table as they were reported by park managers.

QUESTIONNAIRE
Research Park Study

FOR DEVELOPERS OR MANAGERS OF RESEARCH PARKS IN THE UNITED STATES

This questionnaire should take no longer than twenty minutes to complete.

In the questionnaire, the terms "region" and "area" should be interpreted as the Metropolitan Statistical Area (MSA) in which your park is located. (Please call us if you do not know the counties included in your MSA.)

We very much appreciate your willingness to complete this questionnaire. Please return it to us in the enclosed, self–addressed and pre–paid envelope.

If you have any questions, call us, collect if necessary, at (919) 962–3983.

Name of your Park:

Name of person completing this questionnaire:

Phone number of person completing questionnaire:

General Information about the Park

1. Please indicate the organizational status of this park. Is it a: *(check one)*

non–profit private corporation or foundation ☐
unit of a private university ☐
unit of public university ☐
for–profit corporation ☐
unit of state government ☐
university/private developer joint venture ☐
other (please specify: _____) ☐

2.   a. In what year was the park established?   _____
     b. In what year did the first tenant begin operations in the park?   _____

3. How many organizations now are located in this park?   _____

4. Approximately how many people were employed in the park:

This month: _____
2 years ago: _____
5 years ago: _____
10 years ago: _____

5. What is the total acreage of the park?   Acres: _____

6.   a. How many acres have been improved with infrastructure?   Acres: _____
     b. To date, how many acres have been built upon?   Acres: _____

7. What is the primary zoning classification of the park?   _____

8. a. Do you have deed restrictions (covenants) that supersede zoning ordinances?

     NO ☐   YES ☐

<If your answer to #8a is "No," GO TO #9; otherwise, answer #8b – #8f below>

     b. Is there a maximum building/lot ratio?   NO ☐   YES ☐

     c. Is there a minimum lot size?   NO ☐   YES ☐

<If your answer to #8c is "Yes," answer #8d; otherwise, GO TO #8e>
     d. Minimum: _____

     e. Is there a minimum building square footage requirement?   NO ☐   YES ☐

<If your answer to #8e is "Yes," answer #8f; otherwise, GO TO #9>
     f. Minimum: _____

→ 9. Are lots/facilities sold or leased? (*check all that apply*)
    a. Sold ☐
        b. For improved land that is sold, what is the current price per square foot
           or per acre? $__/__

    c. Leased ☐
        d. What percent of organizations in leased spaces are sublessees?   ___%

        e. For improved land that is leased, what is the current average price
           per square foot or per acre? $__/__

        f. What percentage of improved space in the park is leased or subleased?   ____%

        g. What percent of organizations actually occupy the space they own or lease?   ____%

10. Please indicate how your sale and lease prices are determined. Please rank the following statements in their order of importance to you. "1" is most important.
                        We try to assess the current market price ___
                        We try to recover costs ___
                We attempt to meet a "required" rate of return ___
                Other (please specify: _____) ___

Park Operations and Policies

11. Which of the following services do you provide to existing organizations in the park?
    (*Check as many as apply.*)
        Signage ☐
        Land use planning ☐
        Business services, e.g. telephone answering, photocopying ☐
        Natural gas hookups ☐
        Prewired telephone systems ☐
        Liaison with universities ☐
        Liaison with state government agencies ☐
        Liaison with local government agencies ☐
        Security services ☐
        Landscaping and grounds maintenance ☐
        Conference and meeting facilities for professional exchange ☐
        On–site hotel ☐
        Restaurants or other facilities for social exchange ☐
        Construction and maintenance of roadways ☐
        Sewer and water services ☐
        Management consulting ☐
        Other (please specify)_____ ☐

12. Please indicate the relative importance of the following possible park objectives when the park was first proposed. (*Circle one number for each item.*)

	Very Important	Moderately Important	Somewhat Important	Not Important
a. To diversify the region's economic base	1	2	3	4
b. To develop and nurture new businesses	1	2	3	4
c. To capitalize on existing R&D in the region	1	2	3	4
d. To encourage entrepreneurship in the region	1	2	3	4
e. To increase productivity of the economy through innovation	1	2	3	4
f. To increase technology transfer from park businesses	1	2	3	4
g. To enhance university's scientific/ engineering training capability through collaborative research	1	2	3	4
h. To commercialize university–based research	1	2	3	4
i. To enhance the prestige of the affiliated university	1	2	3	4
j. To provide higher paying jobs in the local labor market	1	2	3	4
k. To expand employment opportunities in the area	1	2	3	4
l. To expand employment opportunities for low skilled workers	1	2	3	4
m. To increase employment opportunities for local university graduates	1	2	3	4
n. To maximize profit from the development and sale or lease of park land and facilities	1	2	3	4
o. Other (please specify)	1	2	3	4

13. How have these park objectives changed over time? Please indicate for each goal how the current level of importance compares with the original importance by circling the appropriate number.

	More Important	Same Importance	Less Important	No Longer Important
a. To diversify the region's economic base	1	2	3	4
b. To develop and nurture new businesses	1	2	3	4
c. To capitalize on existing R&D in the region	1	2	3	4
d. To encourage entrepreneurship in the region	1	2	3	4
e. To increase productivity of the economy through innovation	1	2	3	4
f. To increase technology transfer from park businesses	1	2	3	4
g. To enhance university's scientific/ engineering training capability through collaborative research	1	2	3	4
h. To commercialize university–based research	1	2	3	4
i. To enhance the prestige of the affiliated university	1	2	3	4
j. To provide higher paying jobs in the local labor market	1	2	3	4
k. To expand employment opportunities in the area	1	2	3	4
l. To expand employment opportunities for low skilled workers	1	2	3	4
m. To increase employment opportunities for local university graduates	1	2	3	4
n. To maximize profit from the development and sale or lease of park land and facilities	1	2	3	4
o. Other (please specify)	1	2	3	4

14. a. Does your park currently include incubator–type facilities for new, small businesses?

NO ☐   YES ☐

<If you answered "Yes" to #14a, answer #14b; otherwise, answer #14c>

    b. About what proportion of the total floor space in the park is comprised of incubator–type facilities? _____%

    c. Do you plan to have any incubator–type facilities and/or services in the future?

NO ☐   YES ☐

15. Which of the following are <u>permitted</u> uses in the park? (*Check as many as apply.*)

R & D ☐
Prototype manufacturing ☐
Light manufacturing ☐
Heavy manufacturing ☐
Office–general ☐
Corporate or division headquarters ☐
Warehousing ☐
Professional or business services ☐
Residential activities ☐
Retail or consumer services ☐

16. Do you have a limit on the amount of manufacturing performed by park tenants?

NO ☐   YES ☐

<If "Yes", please describe:>

17. In your marketing efforts, do you favor certain types of organizations? (*Please circle one number for each type of business listed.*)

	Strong Preference For	Mild Preference For	Neutral	Mild Preference Against	Strong Preference Against
a. Technology–oriented	1	2	3	4	5
b. R&D	1	2	3	4	5
c. Light manufacturing	1	2	3	4	5
d. Branch plants of national or multinational corporations	1	2	3	4	5
e. Foreign–owned businesses	1	2	3	4	5
f. Start–up businesses	1	2	3	4	5
g. Locally–owned businesses	1	2	3	4	5

18. a. Who owned the land now occupied by the park immediately prior to the park's creation? (*Please check all that apply*)

federal government	☐
state or local government (excluding public univ.)	☐
public university	☐
private university	☐
private individual(s)	☐
real estate developer	☐
non–real estate corporation	☐
other: _____	☐

    b. Please furnish the name(s) of largest landowners prior to park development:

19. Who currently owns the land occupied by the park? (*Please check all that apply.*)
☐ Local universities   Which ones? _____
☐ Local government   Which ones? _____
☐ Individual research firms own their land and facilities
☐ A park development or management firm   Firm name: _____
☐ A consortium of several of the above organizations
        What groups compose this consortium? _____

20. How was land for the park acquired? (*Please check all that apply.*)

    a. University funds   ☐

    b. Donations from private organizations or individuals (non–university)   ☐

    c. State or local government funds   ☐
        i. Was eminent domain used?   YES ☐ NO ☐

    d. Private investor funds   ☐
        i. Was this purchase: (*check only one*)
            1. Piecemeal ☐
            2. One, or only a few large parcels ☐

        ii. What entity were funds paid into?   ☐
            1. a private corporation ☐
            2. a public or quasi–public corporation ☐
            3. other _____

21. What types of government financial assistance (subsidies and grants, or bonds) have been used for park development? (*Please check one*)

    a. No government assistance used ☐

    b. Government subsidies and grants used ☐
        1. What types were used, and by what level of government?
          (*Please circle all that apply*)

	Federal	State	Local
UDAG	1	2	3
CDBG	1	2	3
Subsidized or guaranteed loan	1	2	3
Mortgage subsidy	1	2	3
Land write down	1	2	3
Equity investment	1	2	3

    c. Government bonds ☐
        i. What level(s) of government issued the bonds?   State ☐   Local ☐

        ii. Are bonds tax exempt for: (*Check all that apply*)
            1. Federal taxes   ☐
            2. State taxes   ☐
            3. Local taxes   ☐

        iii. What types of bonds were used? (*Check all that apply*)
            1. General obligation bonds ☐
            2. Revenue bonds ☐
            3. Umbrella bonds ☐
            4. Other (please specify) _____

22. What is the role of venture capital in the development of the park? (*Please check one*)
    a. Venture capital used for equity by the park owner/developer in the development of the park ☐
        b. Venture capital used for equity by individual organizations in the park ☐
            c. Venture capital not used at all ☐

23. We are interested in who takes primary responsibility for PLANNING, FINANCING, and CONSTRUCTION/MAINTENANCE or service DELIVERY of infrastructure and services in the park. For each of these activities, circle as many numbers as apply. NOTE: If private contractors are used, indicate who that contractor reports to.

	Park mgr/ owner	University	State government	Local government	Occupants	N/A
Access roads and highway interchanges						
PLANNING	5	4	3	2	1	0
FINANCING	5	4	3	2	1	0
DELIVERY	5	4	3	2	1	0
Internal roads						
PLANNING	5	4	3	2	1	0
FINANCING	5	4	3	2	1	0
DELIVERY	5	4	3	2	1	0
Public transportation to or within the park						
PLANNING	5	4	3	2	1	0
FINANCING	5	4	3	2	1	0
DELIVERY	5	4	3	2	1	0
Improving air transportation services						
PLANNING	5	4	3	2	1	0
FINANCING	5	4	3	2	1	0
DELIVERY	5	4	3	2	1	0
Improving rail service						
PLANNING	5	4	3	2	1	0
FINANCING	5	4	3	2	1	0
DELIVERY	5	4	3	2	1	0
Water and sewer lines						
PLANNING	5	4	3	2	1	0
FINANCING	5	4	3	2	1	0
DELIVERY	5	4	3	2	1	0
Business recruitment						
PLANNING	5	4	3	2	1	0
FINANCING	5	4	3	2	1	0
DELIVERY	5	4	3	2	1	0
Garbage disposal						
PLANNING	5	4	3	2	1	0
FINANCING	5	4	3	2	1	0
DELIVERY	5	4	3	2	1	0
Industrial waste disposal, other than garbage						
PLANNING	5	4	3	2	1	0
FINANCING	5	4	3	2	1	0
DELIVERY	5	4	3	2	1	0
Fire protection						
PLANNING	5	4	3	2	1	0
FINANCING	5	4	3	2	1	0
DELIVERY	5	4	3	2	1	0
Police/security						
PLANNING	5	4	3	2	1	0
FINANCING	5	4	3	2	1	0
DELIVERY	5	4	3	2	1	0

24. Which of the following items have the state or local governments provided to the park? (*Please check all that apply*)

a. Construction or subsidy of government facilities in the park □
b. Lease or purchase of land or building(s) by public agency □
c. Dedication or gift of publicly–owned land to the park □
d. Property tax reductions, abatements, or exemptions □
e. Special state legislation for the creation of a special tax district for the park □

25. Please indicate the frequency with which the following factors are cited as reasons for organizations' interest in locating in this park. (*Circle one number for each item.*)

a. the presence of a research university in the area   most frequent 1----2----3----4----5 least frequent

b .the specific organizations already located in
    the park   most frequent 1----2----3----4----5 least frequent

c. the concentration of technology–oriented
    organizations already in the region   most frequent 1----2----3----4----5 least frequent

d. the park's land use planning   most frequent 1----2----3----4----5 least frequent

e. the specific facilities, sites, or buildings
    in the park   most frequent 1----2----3----4----5 least frequent

f. the business services and assistance provided
    by park management   most frequent 1----2----3----4----5 least frequent

g. social and cultural amenities of the region   most frequent 1----2----3----4----5 least frequent

h. the business climate of the region   most frequent 1----2----3----4----5 least frequent

i. the amenities of the park as a place to work   most frequent 1----2----3----4----5 least frequent

j. the housing opportunities in the area   most frequent 1----2----3----4----5 least frequent

k. access to business services outside the park
    but nearby   most frequent 1----2----3----4----5 least frequent

l. special incentives offered by state or local
    government   most frequent 1----2----3----4----5 least frequent

m. other (*please specify*:   )   most frequent 1----2----3----4----5 least frequent

26. If there is a research university in the area, we would like to know what prospective tenants value most about it when considering locating in the park. Please indicate the degree of frequency with which each of the following items is mentioned. (*Circle one number for each item.*)

a. Opportunities to have university faculty as
consultants   most frequent 1----2----3----4----5 least frequent

b. Access to recruiting university graduates for
their workforce   most frequent 1----2----3----4----5 least frequent

c. Opportunity to use university facilities including
libraries, labs, or special equipment   most frequent 1----2----3----4----5 least frequent

d. Opportunity for joint research activities between
the business and the university   most frequent 1----2----3----4----5 least frequent

e. Opportunity for scientists or engineers to become
adjunct professors   most frequent 1----2----3----4----5 least frequent

f. Opportunities for employees to take courses at the
university to enhance their training   most frequent 1----2----3----4----5 least frequent

g. Quality-of-life of the area due to the university (e.g.,
cultural activities, intellectual dynamism)   most frequent 1----2----3----4----5 least frequent

h. Other (*please specify* ;   )   most frequent 1----2----3----4----5 least frequent

27. Please indicate the relative importance of each of the following factors as disadvantages for attracting tenants, in comparison to competing parks or regions. (*Please circle one number for each factor.*)

	Major disadvantage	Minor disadvantage	Not a disadvantage
a. No research university nearby	3	2	1
b. Insufficient strength of local university in specific areas of research	3	2	1
c. High cost of sites in the park	3	2	1
d. Inadequate supply of appropriate non-professional labor in area	3	2	1
e. Inadequate supply of appropriate professional labor in area	3	2	1
f. Too high cost-of-living in area, including housing	3	2	1
g. Environmental problems or traffic congestion in area	3	2	1
h. Inadequate public services in area	3	2	1

(continued)

	Major disadvantage	Minor disadvantage	Not a disadvantage
i. No/too small incentives from state or local government	3	2	1
j. Poor access to corporate headquarters	3	2	1
k. Poor access to business's manufacturing plants	3	2	1
l. Inadequate business services in area	3	2	1
m. Park's sites or facilities not suitable	3	2	1
n. Park's restrictions too stringent	3	2	1
o. Other (*please specify*:)	3	2	1

28. Finally, what have been some of the direct <u>and</u> indirect impacts of the park on the region <u>to-date</u>? <u>If</u> the park had <u>not</u> been created, how do you think the following conditions would differ? (*Please circle one number for each item.*)

	Much higher	Somewhat higher	About the same	Somewhat lower	Much lower
a. The area's employment growth rate would be:	5	4	3	2	1
b. New business start-ups in the area would be:	5	4	3	2	1
c. Professionals' compensation level in the area would be:	5	4	3	2	1
d. Non-professionals' pay rate in the area would be:	5	4	3	2	1
e. Local minorities' job opportunities would be:	5	4	3	2	1
f. The area's population would be:	5	4	3	2	1
g. The area's overall quality-of-life would be:	5	4	3	2	1
h. The quality of the area's natural environment would be:	5	4	3	2	1
i. The area's cost-of-living would be:	5	4	3	2	1
j. The concentration of manufacturing in area would be:	5	4	3	2	1
k. Efficiency of land use in area would be:	5	4	3	2	1
l. The quality of the area's universities would be:	5	4	3	2	1
m. The stature of the area's universities would be:	5	4	3	2	1
n. Other (*please specify*:)	5	4	3	2	1

Appendix C.
R&D Organizations in the Research Triangle Park

1. AIRCO Special Gases Electronic Development Facility
2. American Association of Textile Chemists and Colorists
3. BASF Corporation, Agricultural Research Center
4. Battelle Memorial Institute, Applied Statistics Section
5. Becton, Dickinson Research Center
6. Bell Northern Research (BNR)
7. Burroughs Wellcome Co.
8. Cedalion Systems, Inc.
9. Chemical Industry Institute of Toxicology
10. Ciba-Geigy Biotechnology Research
11. Comp-Aid, Inc.
12. CompuChem Laboratories, Inc.
13. Computer Sciences Corp.
14. Data General Corp.
15. Diversified Control Systems, Inc.
16. Du Pont
17. Family Health International
18. FiberLAN, Inc.
19. Glaxo, Inc.
20. GTE South
21. Harris Microelectronics Center
22. H-Three Systems Corp.
23. Institute for Transportation Research and Education
24. Instrument Society of America
25. International Business Machines (IBM)
26. Kobe Development Corp.
27. LITESPEC, Inc.
28. McMahan Electro-Optics, Inc.
29. Microelectronics Center of North Carolina
30. National Center for Health Statistics
31. National Institute of Environmental Health Sciences
32. National Toxicology Program
33. North Carolina Alternative Energy Corporation
34. North Carolina Biotechnology Center
35. Northern Telecom, Inc., Integrated Network Systems
36. NSI Technology Services Corp.
37. Performance Analysis Corp.
38. Program Resources, Inc.
39. QUADTEK, Inc.
40. Radian Corp.
41. Research Triangle Institute
42. Rhone-Poulenc
43. Science and Technology Research Center (STRC)
44. Semiconductor Research Corp.
45. Southeastern Educational Improvement Laboratory
46. Sumitomo Electric Research Triangle, Inc.
47. Troxler Electronic Laboratories, Inc.
48. UAI Technology, Inc.
49. Underwriters Laboratories, Inc.
50. U.S. Department of Agriculture, Forestry Sciences Laboratory
51. U.S. Environmental Protection Agency
52. Wesson, Taylor Wells & Associates, Inc.

Appendix D.
R&D Organizations in the University of
Utah Research Park

1. Advanced Nutritional Research
2. Association of Regional University Pathologists
3. Automated Language Process System
4. Bioactives, Inc.
5. Biomaterials International, Inc.
6. BSD Medical Corporation
7. Cardiopulmonics, Inc.
8. Chem Biochem Research
9. Deseret Research
10. DJH Engineering Center, Inc.
11. Evans and Sutherland Computer Corp.
12. Ford, Bacon and Davis, Inc.
13. Geokinetics
14. Miami Serpentarium Laboratories
15. Nelson Labs
16. Newmont Metallurgical Services
17. Northwest Pipeline Corp.
18. NPI
19. RGB Dynamics
20. Technology Transfer, Inc.
21. Terra Tek, Inc.
22. Thera Tech
23. University of Utah Research Institute
24. Utah Bioresearch, Inc.
25. Utah Geological and Mineral Survey
26. Utah Innovation Center
27. Utah Innovation Foundation
28. VA Region 5 Information System Center
29. Western Institute of Neuropsychiatry
30. Western Systems Coordinating Council

Appendix E.
R&D Organizations in the Stanford Research Park

1. ACCTEX
2. ALZA
3. Beckman Instruments, Inc.
4. California Ear Institute
5. Coherent, Inc.
6. Computer Curriculum
7. The Copper Companies, Inc.
8. Cor Therapeutics
9. DNAX
10. Electric Power Research Institute
11. Fairchild Industries
12. General Instrument Corp.
13. Hewlett-Packard
14. International Business Machines
15. Kaiser Electronics
16. Kodalex
17. Lockheed
18. Molecular Devices
19. Next, Inc.
20. Project Hear
21. Protein Design Labs
22. Scientific Technology, Inc.
23. Sungene
24. Syntex
25. System Control, Inc.
26. Syva Corp.
27. Teledyne
28. U.S. Geological Survey
29. Varian Associates
30. Xerox Corp.
31. Zoecon

Notes

1. We did find some analyses of research parks as real estate ventures, including Weiss, "High-Tech Facilities," and Levitt, *The University/Real Estate Connection*.
2. Miller and Cote, "Growing the Next Silicon Valley."
3. Friedmann, *Regional Development Policy*, p. 20.

Chapter 1

1. Franco, "Key Success Factors for University-Affiliated Research Parks," p. 75, and telephone interview with authors, March 11, 1991. Money, "University Related Research Parks," and Danilov, "How Successful Are Science Parks?" estimate failure rates at least this high.
2. Our definition of high-technology industries combines the criteria of the proportion of R&D inputs as a percentage of sales and the number of scientists and engineers as a percentage of total employment in each industry sector. See Table 1-1 for a list of the industry sectors designated as high-tech. For the classification *advanced services*, see Noyelle, "Advanced Services in the System of Cities."
3. The variation in pretest and posttest periods among parks should not affect the results since (1) each park and its respective control group have the same periods and (2) the employment growth rates are annualized.
4. Cook and Campbell, *Quasi-Experimentation*, chapters 3 and 5.
5. Regional economic development outcomes are also measured as the park managers' perceptions of what difference the research park has meant to its region along a number of dimensions. These data were obtained from a series of items in the park managers' questionnaires. Alternative "objective" outcome measures, such as posttest minus pretest gain in per capita income or rate of new business formation, are discussed in Chapter 3.
6. U.S. Department of Commerce, Bureau of Economic Analysis, *RIMS II*.

Chapter 2

1. For example, see Alexander, "A Park That Reversed a Brain Drain"; Binyon, "Research Parks"; and Fairbanks, "North Carolina Eyes State's 'Golden Goose'." The one "academic" study that is often cited, which contains material on North Carolina's

Research Triangle Park, is Vogel's *Comeback.*

2. Franco, "Key Success Factors for University-Affiliated Research Parks."

3. See Holland, *Capital versus the Regions*; Rees and Stafford, "A Review of Regional Growth and Industrial Location Theory"; and Gore, *Regions in Question*. Gore distinguishes Schumpeterian theories that "focus on the way in which entrepreneurial behavior and the act of innovation affect the dynamics of capitalist economies" (p. 86) from Perrouxian theories that stress "a range of mechanisms through which a propulsive unit could induce growth in many other parts of the economy" (p. 86).

4. Pred, "The Interurban Transmission of Growth in Advanced Economies"; Perroux, "Economic Space, Theory, and Applications"; Boudeville, *Problems of Regional Economic Planning*; Hansen, "Development of Pole Theory in a Regional Context"; Hirschman, *The Strategy of Economic Development*; Myrdal, *Economic Theory in Underdeveloped Regions*. For a discussion, see Darwent, "Growth Poles and Growth Centers in Regional Planning."

5. Berry, *Growth Centers in the American Urban System* and "Hierarchical Diffusion"; Lasuén, "Urbanization and Development" and "Multi-Regional Economic Development"; Thwaites, Edwards, and Gibbs, *Interregional Diffusion of Product Innovations in Great Britain*; Thwaites, Oakey, and Nash, "Technological Change and Regional Development"; Rees, Briggs, and Hicks, "New Technology in the United States' Machinery Industry"; Junne, "Der Strukturpolitische Wettlauf zwischen den Kapitalistische Industrielandern"; Schamp, "Technology Parks and Interregional Competition in the Federal Republic of Germany."

6. See Pred, "The Interurban Transmission of Growth in Advanced Economies"; Gilmour, "External Economies of Scale"; Britton, "Environmental Adaptation of Industrial Plants"; Erickson, "The Spatial Pattern of Income Generation"; and Moseley, *Growth Centres in Spatial Planning.*

7. Higgins, "From Growth Poles to Systems of Interaction in Space."

8. It is important to note, however, that a concentration of growth in the center could serve to exacerbate regional inequalities. A third possibility is that little growth is induced by research parks anywhere. Then, we can question involvement by any level of government in park development.

9. The particular amenities deemed most important have been studied by Premus, *Location of High Technology Firms and Regional Economic Development*; Malizia, "The Locational Attractiveness of the Southeast to High-Technology Manufacturers"; and Markusen, Hall, and Glasmeier, *High Tech America*. Depth and breadth of business services, and urbanization economies, more generally, may be less important locational factors for R&D–based organizations than the requirement growth center doctrine places on growth centers when R&D is not the propulsive activity.

10. The term *product cycle* comes from Vernon, "International Investment." Vernon argues that industries go through three stages (innovation, growth, and maturation), each with locational implications. For recent applications of this theory to regional development, see Glasmeier, *The Making of High Tech Regions*, and Markusen, *Profit*

Cycles, Oligopoly, and Regional Development.

11. Induced economic growth stemming from enhanced productivity in existing enterprises through technology transfer and innovation from propulsive industries in growth centers is not considered important in growth center doctrine.

12. See Krist, "Innovation Centres." For a more complete discussion of the dichotomy between regional economic (exogenous) and entrepreneurial (indigenous) approaches, see Allen and Hayward, "The Role of New Venture Formation/Entrepreneurship."

13. See R. N. Cox, "Lessons from 30 Years of Science Parks," and Thomas, "Regional Economic Development and the Role of Innovation and Technical Change."

14. Andersson, "Creativity and Regional Development," p. 19; Scott and Storper, "High Technology Industry and Regional Development."

15. Bollinger, Hope and Utterback, "A Review of Literature and Hypotheses"; U.S. Congress, Office of Technology Assessment, *Technology, Innovation, and Regional Economic Development*; R. N. Cox, "Lessons from 30 Years of Science Parks" (quotations, pp. 17, 20); Cooper and Komives, *Entrepreneurship*, pp. 108–25; Harris, "Establishing High-Technology Enterprises in Metropolitan Areas." See Shapero, "Entrepreneurship," on the density of innovators, and R. N. Cox, "Lessons from 30 Years of Science Parks," p. 20, on the notion of "fertilizers for the seedbed."

16. Schamp, "Technology Parks and Interregional Competition in the Federal Republic of Germany," pp. 129, 131.

17. Malecki, "Dimensions of R&D Location."

18. A similar conclusion was reached by Oakey, *High Technology Industry and Industrial Location*, in a study of the scientific instruments industry in Great Britain.

19. Rees, "The Diffusion of New Production Technology."

20. The direction of causality could be in the opposite direction, so that the R&D facility is located close to the manufacturing facility. Close linkage between product engineers and marketing personnel, who are usually based at corporate headquarters, may mean that the R&D facility is located near headquarters.

21. This, too, is confirmed by Oakey, *High Technology Industry and Industrial Location*.

22. Darwent, "Growth Poles and Growth Centers in Regional Planning."

23. Malecki, "Recent Trends in the Location of Industrial Research and Development"; Browning, *How to Select a Business Site*.

24. For example, Glasmeier, *The Making of High Tech Regions*, p. 29, and Malecki, "Recent Trends in the Location of Industrial Research and Development."

25. See McDonough, "Suburban Industrial Park Development."

26. There is a growing literature on the university-research park relationship. See, for example, O'Connor, "Mutual Benefits Boost University-Related Research Parks"; Hull, "The University and Economic Development"; Levitt, *The University/Real Estate Connection*; Glazer, "Business Takes Root in University Parks"; and Steinnes, "Evaluating the University's Evolving Economic Development Policy," pp. 214–26.

27. There is some debate about which university resources are most important in inducing other R&D activity. It is reasonable to assume that small- and medium-sized technology-oriented firms would take advantage of universities for specialized facilities and faculty consulting. Yet, Sirbu et al., *The Formation of a Technology Oriented Complex*, suggests that new technology-oriented firms are less likely to have linkages with universities because of "cultural differences" and less chance of matching the applied research needs of the firm with university research activities. Williams, "University-Industry Interactions," discusses the difficulties of matching the applied research needs of small- and medium-sized businesses with university research activities. The accessibility to a specialized entry-level labor force (i.e., university graduates) may well be the more important university resource for both large and small technology-oriented enterprises in the region.

28. Franco, "Key Success Factors for University-Affiliated Research Parks."

29. Malecki, "Dimensions of R&D Location"; Stöhr, "The Spatial Dimension of Technology Policy."

30. Norton and Rees, "The Product Cycle and the Spatial Decentralization of American Manufacturing."

Chapter 3

1. In the case of small new firms, often in higher-density, urban areas, parks provide services and facilities at lower cost to the individual business because the concentration of businesses with similar needs creates economies of scale. This, of course, is the concept behind incubators.

2. In some cases, parks are governed by occupants who decide how to spend at least a portion of the park corporation's funds.

3. Approximately 10 percent of our sample of park developers and 34 percent of our sample of park occupants have used venture capital to finance their start-up operations.

4. We exclude from the discussion here the role of institutions that provide debt finance.

5. See Goldstein and Luger, "Research Parks."

6. This finding is based on our interviews with state and local officials in the three case study areas.

7. The parks to which this applies are state- or municipally owned or part of a public university.

8. Another important state government activity is the creation of special park districts, protecting the parks from municipal annexation. Park occupants then pay no city taxes. They either purchase services from nearby municipalities or have the park manager provide them. This legislation has been passed for approximately 10 percent of the parks surveyed in our study.

9. These findings are based on our interviews with officials at universities affiliated with the three case study research parks.

10. See Oakey, *High Tech Industry and Industrial Location*; Markusen, Hall, and Glasmeier, *High Tech America*; and Malecki, "Dimensions of R&D Location" and "The R&D Location Decision."

Chapter 4

1. In the course of the research, we generated the master list of 116 research parks from a roster of parks belonging to the Association of University-Related Research Parks, references to parks in trade journals and the literature, and referrals by park managers and developers. A reconnaissance survey of those parks revealed that approximately 30 were not currently active. A more detailed questionnaire was sent to the managers of approximately 90 parks, 76 of which responded. Four of the 76 parks were officially open but did not have tenants when the questionnaire was completed. The 14 parks that failed to respond did not have many employees. A copy of our park manager questionnaire is in Appendix B.

2. Note that percentages are not provided here since parks often use multiple sources of funds to obtain land.

3. For comparative purposes, in 1989, according to the National Association of Realtors, the national average price per square foot of leased class A suburban office space was $15 to $21, and the national average rental rate for prime industrial space in suburban areas was $4.70 per square foot.

4. In our sample of seventy-two parks, 94.4 percent prohibit heavy manufacturing, 86.1 percent prohibit residential uses, 76.4 percent outlaw warehousing, 58.3 percent forbid retail and consumer services, and 30.6 percent prohibit light manufacturing.

5. The thirty parks with minimum lot size requirements are distributed as follows: less than 1.1 acre (9), 1.1 to 2.5 acres (9), 2.6 to 5 acres (10), and more than 5 acres (2).

6. See, for example, Allesch and Fiedler, *Management of Science Parks and Innovation Centers*; Gibb, *Science Parks and Innovation Centres*; and Sternberg, *Technologie- und Gründerzentren*.

7. Wilson, *The Research Triangle of North Carolina*.

8. See Cook and Campbell, *Quasi-Experimentation*, for an overview of this class of research designs.

9. The reader should refer to Chapter 3, where this measure is presented formally.

10. The 1984 cutoff was necessary because of the forward lag.

11. As we have demonstrated in earlier chapters, the employment effect often comes as much from households spending payroll, business purchasing, and construction multipliers as from direct employment in park businesses.

12. The classification of universities comes from the Carnegie Foundation for Higher Education. Type I and Type II are research universities. The designation of a research university is based on the annual amount of federal government contracts and grants for research and the annual number of Ph.D. degrees awarded. Type I universities conduct a greater amount of federally funded research and grant a larger number of doctorates than

Type II universities. "Doctoral granting universities" is another classification of institutions that award Ph.D.s but do not meet the threshold for federal research contracts and grants to be classified as a Type II research university.

13. Several different mathematical forms of the regression model were used. We experimented with different indicators of success from Table 4-5, and other measures discussed in Chapter 3, as the dependent variable, and with the characteristics listed in Table 4.7, and other explanatory variables, as regressands. We estimated the linear OLS model using a continuous dependent variable, and with nonlinear logit and probit models, using a dichotomous regressor. Complete model specifications and results can be obtained from the authors.

14. If "DIFF" in Table 4-5 is positive, then the particular park is designated as successful and assigned a value of one. If "DIFF" is zero or negative, then the park is designated as not successful and assigned a value of zero. The details of the logit model are available in standard econometric texts. See, for example, Manski and McFadden, *Structural Analysis of Discrete Data with Econometric Applications*. Logit estimation allows one to interpret coefficients probabilistically when the dependent variable is discrete. We also ran probit models, which differ in their error structures and optimization algorithms, and obtained similar results.

15. If the number of parks is r_j, the size of the risk set is $n_j - C_j/2$, where C_j is the number of censored observations. The proportion of parks exiting (failing) is denoted as $q_j = m_j r_j$, where m_j is the number of exits and r_j is the risk set. The cumulative hazard rate is $x_j = 2q_j/h(2 - q_j)$, where h is the interval width. For a complete discussion of this technique, see Kalbfleisch and Prentice, *The Statistical Analysis of Failure Time Data*; Cox and Oakes, *Analysis of Survival Data*; and Gross and Clark, *Survival Distributions*.

16. The necessary regression model employs maximum likelihood estimation similar to the logit framework used above. The basic function is $h(t, w) = h(o, w)e^{xw}$ where $h(o, w)$ is the baseline hazard rate at time o, where w is the vector of covariates. See D. R. Cox, "Regression Models and Life Tables."

Chapter 5

1. For a complete history of the Research Triangle Park, see Sellars, "An Oral History of the Research Triangle Park." Other useful historical references include Wilson, *The Research Triangle of North Carolina*, and Hamilton, "The Research Triangle of North Carolina." The brief historical sketch in this case study borrows from those works and from Little, "Research Triangle Park," and uses information gained in interviews with RTP officials.

2. Little, "Research Triangle Park."

3. Ibid.

4. Ibid.

5. According to one popular but unverified story, the NIEHS facilities in the park

were a payoff to Terry Sanford for making a nominating speech at the 1960 Democratic National Convention for John F. Kennedy.

6. Although few new enterprises have located in the park, several of the *existing* businesses have made large expansions in their facilities and workforces during this period.

7. Population data are from the U.S. Bureau of the Census, *1960 Census of Population*. Employment data are from the U.S. Bureau of the Census, *County Business Patterns*.

8. Data are from the National Planning Association, *Regional Economic Projection Series, 1987*.

9. Data are from James O. Roberson, president of the Research Triangle Foundation, letter to the authors, August 24, 1989.

10. Data are from the U.S. Bureau of the Census, *County Business Patterns*.

11. Data on educational attainment are from the U.S. Bureau of the Census, *1960 Census of Population*. Although the proportion of persons twenty-five years and older who did not finish high school in the Raleigh-Durham area in 1960 (56.6 percent) was slightly below that for the United States as a whole (58.9 percent), we still interpret the area's figure as unusually high for metropolitan areas of the country.

12. Southern Growth Policies Board, *Halfway Home and a Long Way to Go*.

13. Data are based on telephone interviews with admissions offices and registrars in appropriate academic units at the three universities.

14. Luebke, "Corporate Conservativism and Government Moderation in North Carolina"; Goldman and Luebke, "Corporate Capital Moves South"; Luebke, Peters, and Wilson, "The Political Economy of Microelectronics in North Carolina." These articles discuss the differences between traditionalists and modernizers in the political economy of the New South.

15. Data are from interviews with Elizabeth Johnson Aycock, secretary of the Research Triangle Foundation, who has served on its staff since its founding, and Dr. William Little, professor in the University of North Carolina's chemistry department, an early promoter of the park and a member of the Research Triangle Foundation's board of directors. Interviews were conducted in the summer of 1988 in the Research Triangle Park and Chapel Hill, respectively. The business leaders referred to in the text include Robert M. Hanes and Archie K. Davis, successive chairmen of Wachovia Bank and Trust Co. in Winston-Salem, and Romeo H. Guest, an industrialist from Greensboro, N.C., who is widely credited with coining the name *Research Triangle*. Until recently Davis served as chairman of the Research Triangle Foundation's board, a post he held for many years.

16. A footprint is the ratio of the amount of land area on which there are buildings to the total land area of the parcel. Much of the remaining parcel land area is devoted to grounds plus outdoor parking facilities. This is different than the term *floor area ratio*, which is the ratio of total square footage in buildings to parcel land area.

17. In fact, RTP owners and tenants, like *all* non–Durham city users of water and sewer services, pay a premium of about 50 percent.

18. Troxler Electronics Laboratories, Inc., which produces scientific instrumentation, is a counterexample. Troxler was founded in Raleigh, N.C., and then moved to its present location in the park in 1974.

19. This section describes only R&D organizations in the park. Service organizations and office facilities located at RTP are not included.

20. These estimates are based on the responses from the CEOs of park organizations to the question of whether the organization would have located in the region if the research park had not existed. Possible responses were (1) very likely, (2) likely, (3) maybe, (4) unlikely, and (5) very unlikely. The estimate of the percentage of organizations (and the employment in these organizations) that would have located in the region was calculated by assigning the following probabilities to the response: very likely = 0.9, likely = 0.7, maybe = 0.5, and unlikely and very unlikely = 0.0.

The estimate of the number of organizations that would not have located in the region but for the park was calculated as the sum of the probabilities over all respondents. We then inflated this number for the full population of organizations to include non-respondents. The employment estimate was made by multiplying the probability times the employment in each organization. We inflated the result in the same way to include nonrespondents.

21. The estimate for the percentage of firms in the sample that would not have located in the region except for the park is based on the same procedure used for park organizations. The estimate of indirect employment stimulated by the park was made like that for employment inside the park. The only difference is due to the fact that a sample was used rather than the full population. The size of the out-of-park population in employment terms was calculated by subtracting all employment in the park from the growth of employment in the region's high-tech sector between 1960 and 1987. Percentages and probabilities derived from the sample were then applied to this population.

22. Some portion of the high-tech businesses and employment included in our estimate of that induced to the park by localization economies may have located in the region as suppliers of inputs to park organizations. In these cases, this employment impact was double counted. We believe, however, that it was small.

23. These estimates were made by applying data from the U.S. Department of Commerce, Bureau of Economic Analysis, regional input-output tables (*RIMS II*), for North Carolina to the information obtained from the park organizations and sample of out-of-park high-tech businesses. The steps in the estimation of jobs generated by the regional income multiplier were as follows:

E_1 = number of jobs inside the park "caused" by the park = 18,900.

E_2 = number of jobs in the high-tech sector outside the park stimulated by the park = 1,240.

L = average annual salary of employees from E_1 and E_2 = \$40,000.

C_1 = increment to regional payroll from E_1 and E_2 = $(E_1 + E_2) \times L$ = \$805.6 million.

P = estimate of purchases from local businesses (in \$ millions) by park organizations and out-of-park high-tech businesses "caused" by the park = $620.9 million (from responses to questionnaires).

p = proportion of purchases from local businesses paid to labor = 0.30 (from regional input-output table).

C_2 = increment to regional payroll from local business purchases = $186 million.

C = $C_1 + C_2$ (total increment to regional payroll) = $991.6 million.

F = employment multiplier of households (from regional input-output table) × 0.9 leakage factor to rest of state) = 24.3 jobs/$ million of earnings.

E_3 = total number of jobs created through regional income multiplier = C × F = 24,095.

The estimate of the number of jobs generated by local business purchases by businesses induced by the park is:

$E_4 = C_2/H$, where H is the average annual payroll/employee in sectors providing inputs to park organizations and businesses induced by the park ($25,000 in North Carolina) = 7,440.

24. The impact of the park on *real* income levels is not as clear. Average housing costs in the area have gone up substantially, no doubt as a partial result of the influx of higher-income professionals moving to the area because of the park. Though this dimension lies outside the scope of our study, we note that for some nonaffluent sectors of the population, including renters, the increase in housing costs may be offsetting or surpassing any increased earnings indirectly induced by the park.

25. Actually, we would need wage and salary level data for employment stimulated by the park as well as for the workforce in the park. It was not possible to obtain this information due to confidentiality restrictions and resource constraints.

26. The GINI coefficient measures the degree of income inequality on a scale of 0 to 1, where 0 means perfect equality and 1 is the maximum degree of inequality.

27. The data on the sex and race composition of the park's workforce come from responses to the questionnaire supplied by each organization. If there is a systematic bias in these responses, it is probably in the direction of overestimating these proportions.

28. The sample of faculty was drawn from the schools or departments in the following subject areas: agriculture, architecture, business, computer and information sciences, engineering, foreign languages, health (including medicine), life sciences, mathematics and statistics, the physical sciences, and the social sciences.

29. Patenaude, "High-Tech Businesses in Durham County," obtained similar results from a sample of high-tech businesses that had recently located in Durham County.

30. See Shapero, "Entrepreneurship." The Silicon Valley phenomenon and the rather unhealthy economic conditions in most of North Carolina thirty years ago are two different counterexamples to this line of argument.

31. See n. 14 above.

Chapter 6

1. These figures include University of Utah academic and administrative departments, as well as business service organizations (e.g., hotel, bank, and day-care facility), located in the park. About thirty R&D-oriented organizations in the park (excluding university academic departments) account for about 3,800 employees.

2. Charles Evans, interview with one of the authors, Salt Lake City, July 8, 1988.

3. Population and employment figures are from the U.S. Bureau of the Census, *State and Metropolitan Data Book, 1986,* and the National Planning Association, *Regional Economic Projections Series, 1987.*

4. Reno, Nev., with a population of 212,000 (1984 U.S. Census estimate), lies 520 miles west of Salt Lake City on Interstate 80 going toward San Francisco.

5. U.S. Bureau of the Census, *County Business Patterns.* The definition of *high-technology sector* is based on Vinson and Harrington, *Defining "High Technology" Industries in Massachusetts,* and Markusen, Hall, and Glasmeier, *High Tech America.* See Table 1-1 for a list of the specific SIC categories.

6. Brown, "Locally-Grown High Technology Business Development."

7. Dr. Norman Brown, director, University of Utah Technology Transfer Office, interview with one of the authors, Salt Lake City, July 12, 1988.

8. U.S. Bureau of the Census, *1970 Census of Population.*

9. The park director, Charles Evans, mentioned that some competing industrial and office parks in the Salt Lake City region mistakenly thought that rents in the park were being subsidized by the university or by state or local government. Otherwise, they argued, rents should be even higher. (Interview with one of the authors, Salt Lake City, July 8, 1988.)

10. Charles Evans, interview with one of the authors, Salt Lake City, July 12, 1988.

11. University departments or research units located in the park are not included here.

12. The validity of claims of causality varies among the different types of putative impacts (see Chapter 2). We qualify causal claims in the text when available evidence only supports claims of association based on correlations.

13. These estimates are based on the responses of the CEOs of park organizations to the question of whether the organization would have located in the region if the research park had not existed. See Chapter 5, n. 20, for details.

14. The sample was a stratified random sample of businesses in selected SIC categories defined as high-tech sectors.

15. These estimates were made by applying data from the 1986 U.S. Department of Commerce, Bureau of Economic Analysis, regional input-output tables (*RIMS II*) for Utah to the information obtained from the park organizations and the sample of out-of-park high-tech businesses. The steps in the estimation of jobs generated by the regional earnings multiplier were as follows:

E_1 = number of jobs inside the park "caused" by the park = 800.

E_2 = number of jobs in the high-tech sector outside the park induced by the creation of the park = 1,000.

L = average annual salary of employees from E_1 and E_2 = \$30,000.

C_1 = increment to regional payroll from E_1 and E_2 = $(E_1 + E_2) \times L$ = \$54.0 million.

p = proportion of purchases from local businesses paid to labor = 0.30 (from regional input-output table).

C_2 = increment to regional payroll from local business purchases = \$16.48 million.

C = $C_1 + C_2$ (total increment to regional payroll) = \$70.48 million.

F = employment multiplier of households (from regional input-output tables) = 4.5.

E_3 = total number of jobs created through regional income multiplier = $C \times F$ = 1,725.

The estimate of the number of jobs generated by local business purchases by businesses induced by the park is:

E_4 = C_2/H, where H is the average annual payroll/employee in sectors providing inputs to businesses induced by the park (\$20,000 in Utah) = 825.

16. The personal income data are from the U.S. Department of Commerce, Bureau of Economic Analysis, *Local Area Personal Income*, data tapes.

17. See, for instance, Saxenian, *Silicon Chips and Spatial Structure*, and Sampson, "Employment and Earnings in the Semiconductor Electronics Industry." Kuttner, "The Declining Middle," and Bluestone, Harrison, and Gorham, "Storm Clouds on the Horizon," make a more general argument about the "declining middle" in which the growth of the high-tech sector is implicated.

18. Census data were taken from the U.S. Bureau of the Census, *1970* and *1980 Census of Population*. See Chapter 5, n. 26, for a definition of the GINI coefficient.

19. Pretests of questionnaires sent to businesses inside and outside of research parks indicated decisively that these questions were considered both highly sensitive and difficult to answer and thus threatened our ability to achieve an acceptable response rate.

20. U.S. Bureau of the Census, *1970* and *1980 Census of Population*.

21. An even more dramatic improvement in the relative unemployment rate for minorities occurred between 1970 and 1980. Yet because the area has such a small proportion and number of minorities, and because many of these are Asian-Americans rather than blacks, we do not wish to make inferences about changes in the labor market situation of minorities in the area.

Chapter 7

1. There was no grand opening but an uncertain beginning in 1951, when Varian Associates broke ground for a facility on Stanford land. Until 1953, only forty acres were used for industrial purposes. In some documents, 1953 is given as the opening year. See Lowood, "From Steeples of Excellence to Silicon Valley." Two other parks opened in

the first two years of the decade—Swearingen (Oklahoma) in 1950 and Cornell (New York) in 1951—but they have not become as prominent as Stanford.

2. Much of the material that follows is based on an extensive interview with Alf Brandin, former vice-president of Stanford University and one of the key figures responsible for the park. One of the authors interviewed Brandin on July 7, 1988, at his home on the Stanford campus. A follow-up interview was conducted by telephone on August 28, 1989, after Brandin had read an earlier draft of this chapter.

3. The endowment data are from the *San Francisco Examiner*, January 26, 1953. Prices were deflated using the consumer price index.

4. Lowood, "From Steeples of Excellence to Silicon Valley."

5. Quotations from Gibson, *Stanford Research Institute*, pp. 14–15.

6. This point was made by Thomas Ford, current Stanford trustee and former associate of Alf Brandin, when interviewed by one of the authors in Stanford, Calif., on July 8, 1988.

7. This activity had been brought to Stanford planners' attention because both Stanford and the University of Washington were members of the Association of Western Universities and Colleges.

8. "What's Happened Down on the Farm?" Of approximately 8,800 acres of Stanford-owned land, roughly 1,000 acres were in campus development and approximately 3,300 were reserved for future academic use. That left 4,500 acres for uses to be determined.

9. The secondary sources we have examined are not precise about the exact size of the first parcel, only that it was more than forty and less than fifty acres. According to Brandin, it was exactly fifty acres.

10. Lowood, "From Steeples of Excellence to Silicon Valley," p. 11. Brandin says that because Stanford could not file a subdivision map, it was required to develop larger than one-acre parcels on this early land.

11. Murphy, Speech to the Second International Conference of the Association of University-Related Research Parks.

12. The following information on Varian Associates borrows heavily from Lowood, "From Steeples of Excellence to Silicon Valley."

13. Ibid., p. 12. Several faculty members, including Terman, sat on Varian's board of directors, and several former faculty members were principals at Varian.

14. There is a discrepancy in the records we have reviewed. One archival document, dated July 1, 1960, indicates that Varian leased 16.983 acres in the park in 1951.

15. *San Francisco Examiner*, January 26, 1953; Brandin, telephone interview with one of the authors, August 28, 1989. Varian's $1.5 million loan from the Reconstruction Finance Administration was part of post–World War II legislation "that benefited rapidly expanding technology-based companies in areas deemed vital to national defense." Lowood, "From Steeples of Excellence to Silicon Valley," p. 12.

16. Brandin claims that Houghton Mifflin and Scott, Foresman—both publishing firms—only moved to the park when they were confident that there would be a small agglomeration of publishing companies there.

17. Brandin and William Zaner, current city manager of Palo Alto, interviews with one of the authors, Palo Alto, July 10, 1988.

18. The idea of industrial parks dates back to Howard, *The Garden City Movement*. In a letter to one of the authors dated August 30, 1989, Lowood said: "Supposedly, Lewis Mumford was invited as a consultant c. 1946. I have not seen the documents. This involved the Planning Office, not Brandin's, but surely Mumford would have introduced the 'Park' concept."

19. Quoted in Lowood, "From Steeples of Excellence to Silicon Valley," p. 11. Brandin disputes this point. He says that by 1953, the park had been expanded twice and there was general satisfaction with the development that had occurred.

20. Stanford University *Faculty-Staff Newsletter*, March 20, 1960.

21. "What's Happened Down on the Farm?"; Thomas Ford and James Balderston, interviews with one of the authors, Palo Alto, July 8, 1988.

22. Several writers trace the growth of Silicon Valley from Stanford Research Park. See Saxenian, "The Genesis of Silicon Valley"; Lowood, "From Steeples of Excellence to Silicon Valley"; and Murphy, Speech to the Second International Conference of the Association of University-Related Research Parks.

23. U.S. Bureau of the Census, *County Business Patterns*. The industries designated as high-tech or R&D-intensive are listed in Table 1-1.

24. According to Alf Brandin, university planners considered incorporating the park into its own municipality, as Woodside did, and as was done with Menlo Park. Ultimately, Stanford decided that it did not want responsibility for infrastructure and services and felt that a new city on Stanford land would change the nature of the university.

25. *Palo Alto Times*, January 24, December 5, 1956. In an interview with one of the authors (Stanford, Calif., July 11, 1988), Zera Murphy questioned the accuracy of these accounts.

26. *Los Altos News*, January 14, 1955.

27. There is a discrepancy in the material we have reviewed. The initial zoning may have required a 20 percent footprint and a 30 percent open space set-aside, so that these requirements represent a later tightening.

28. Zera Murphy, interviews with one of the authors in Stanford, Calif., July 9, 1988, and by telephone, September 18, 1989. The contracts are consistent with California's Williamson Act.

29. Emily Renzel, member of Palo Alto City Council, telephone interview with one of the authors, July 9, 1988. See also California Department of Health Services, *Stanford Research Park Sites*. Renzel says the city spends $250,000 a year on hazardous waste problems. Stanford officials contest that figure. The city's costs would be to oversee compliance, and those costs should be recoverable via user fees.

30. Murphy, interview with one of the authors, Stanford, Calif., July 9, 1988.

31. Thomas Ford, interview with one of the authors, Stanford, Calif., July 8, 1988.

32. Murphy, interview with one of the authors, Stanford, Calif., July 10, 1988.

33. Brandin, telephone interview with one of the authors, August 28, 1989.

34. A route similar to the one that was built is shown on a university master plan from

the early 1950s. The state paid $5.5 million for condemned Stanford land.

35. Murphy, interview with one of the authors, Stanford, Calif., July 9, 1988. According to William Zaner, the city obtains power in bulk from the federal government under a long-term arrangement that was struck when the government had excess power that was sold at a deep discount. Interview with one of the authors, Palo Alto, July 10, 1988. Brandin says that the power is produced by Pacific Gas and Electric. Interview with one of the authors, Stanford, Calif., July 7, 1980. We asked business representatives what they thought Stanford should provide. Twenty-two out of twenty-four respondents said "land use planning and growth management," fifteen out of twenty-four said "liaison with Stanford," and fifteen out of twenty-three said "amenities."

36. Appendix E contains a list of R&D organizations in the park as of July 1988. Twenty-seven of these organizations responded to a mail survey or submitted to a face-to-face interview during the summer of 1988. See Chapter 1 for more details on the interviewing and mail survey strategy.

37. Thirty-three percent of the respondents said that basic research is the most important function, 22.2 percent indicated that routine production is, and 15 percent said that applied research or product development are most important. The rank of functions in the text is for the aggregation of first, second, and third most important functions performed in the park.

38. This explanation was confirmed in survey responses from out-of-park businesses. Almost half of those businesses pin the difficulty in obtaining semiskilled and unskilled workers on the low basic skill level of the workforce, while 37.1 percent cited an insufficient number of trained graduates. De Anza and Foothill community colleges graduate thousands of students each year. It is possible that the rate of out-migration of these graduates exceeds the rate of in-migration of similarly trained individuals, owing to the high cost of living in the region.

39. For example, see Feller, "University-Industry R&D Relationships"; National Science Board, *University-Industry Research Relationships*; and Steinnes, "Evaluating the University's Evolving Economic Development Policy."

40. The following discussion of the "university connection" is based on questionnaire responses from (1) Robert L. Byer, vice-provost and dean of research; R. H. Eustis, associate dean of the School of Engineering; Kent Peterson, associate vice-president for business affairs; Robert Freelen, vice-president for public affairs; and Neils Reimer and associates in the Office of Technology Licensing; and (2) thirty-four faculty members from Stanford. More detailed discussion of the impact of the park on the university is contained in Chapter 8.

41. Based on survey responses from Stanford faculty, this estimate may be seven percentage points too high.

42. The estimation procedure described in Chapter 5, n. 20, was used here. Questionnaires were sent to approximately five hundred businesses in Santa Clara and San Mateo counties, belonging to twenty-six three-digit SIC groups, chosen for reasons explained in Chapter 1. Eighty-nine completed questionnaires were returned, for a response rate

of 18 percent. Responses are representative of the full sample, and the response rate is typical for mail surveys to private businesses. See Chapter 5, n. 21, for details on the estimation procedure used to calculate these numbers.

43. The payroll for the 1,700 jobs in the park and the 22,000 jobs outside the park, which probably would not exist if the park had not been established, is estimated to be approximately $1,560,000,000. The increment to the regional payroll from local business purchases is estimated to be $202,445,600. The methodology used for these and other computations below is described in Chapter 1.

In 1988, park businesses spent about $835 million on nonlabor inputs. Of this, approximately $313 million was spent in the region and $286 million as elsewhere in the state. Of the amount spent in the region, approximately $15 million would not have been spent if the park had not existed. Of the $1.2 billion spent on nonlabor inputs by out-of-park businesses in 1988, we estimate that $650 million was spent within the metropolitan area. Of that, almost $14 million was spent by businesses that would not exist if the park had not been developed.

44. See Chapter 5, n. 26, for a description of the GINI coefficient.

45. Saxenian, "The Urban Contradictions of Silicon Valley."

46. In 1980, the Asian and Pacific Islander population in California was 1,253,818, or 5.3 percent of the total population. In the same year, the state's black population was 7.7 percent of the total. The Asian and Pacific Islander population in North Carolina was 0.4 percent of the total, and in Utah, it was 1 percent of the population. For the entire United States, the percentage of Asian and Pacific Islanders was 1.5 percent. Census data also reveal that in all regions of the United States, per capita personal income was substantially higher for Asians and Pacific Islanders than it was for blacks.

47. In 1980, the unemployment rate in California was 4.6 percent for Asians and Pacific Islanders and 6.5 percent for blacks, so this paragraph understates the labor market problems in the region for blacks, since it refers to all minorities.

48. Housing price data are from the U.S. Bureau of the Census, *1950* and *1980 Census of Housing*. Data were deflated using the home ownership component of the consumer price index.

49. This estimate was made by extrapolating the number of spin-offs identified in the out-of-park business surveys to the full population of high-tech businesses in the region. The estimate is consistent with the results of a study by Charles Krenz ("Silicon Valley Spin-Offs") of the Stanford Business School. Krenz traced at least 107 spin-offs in the vicinity of Palo Alto that were begun by Stanford faculty.

50. Niels Reimers, interview with one of the authors, Stanford, Calif., July 9, 1988.

51. Henry Lowood, letter to one of the authors, August 30, 1989. Interviews with Stanford officials also revealed concern about more and more general office activity, as opposed to R&D activity, within park establishments.

Chapter 8

1. See, for example, von Blum, *Stillborn Education*; Geiger, *To Advance Knowledge*; and Klausner, "Dilemmas of Usefulness."

2. Feller, "Universities as Engines of R&D-Based Economic Growth." The empirical basis for each of these beliefs—that university-industry partnerships enhance national productivity, increase revenues, and contribute to regional economic development—is discussed more fully in Goldstein and Luger, "Universities and Regional Economic Development in the 1990s."

3. Small business assistance centers provide technical and managerial assistance to small businesses but in a more formal setting than other assistance operations. SBACs take different forms, but most are physical facilities (offices) that house programs to provide university-based expertise to managers of existing small businesses and to entrepreneurs planning to start their own businesses.

4. Inflating this for the universe of Type I-IV institutions produces the estimate above that approximately one hundred universities are involved with research parks.

5. The average annual growth rate in employment from 1976 to 1985 was 9.6 percent for counties containing for-profit parks, while it was 6.1 percent for counties with parks owned by public universities, 5.6 percent for counties with parks operated by state government, 5.3 percent for counties with parks operated by private universities, 5.1 percent for counties with parks operated by nonprofit corporations or foundations, and 2.8 percent for counties with parks operated as joint ventures.

6. Nelson, "Institutions Supporting Technical Advances in Industry"; Bania, Eberts, and Fogarty, "The Role of Technical Capital in Regional Growth"; Jaffe, "Technological Opportunity and Spillovers of R&D" and "Academic Research with Real Effects"; Florax and Folmer, "Regional Economic Effects of Universities," p. 28.

7. Counties containing research universities may have grown relatively slowly in the past decade because the types of businesses often found in their economic bases—for example, R&D and high-tech manufacturing—are relatively slow growing *in terms of employment* but more robust in terms of value added and income generation. In fact, when growth rates in per capita personal income are compared for counties with and without Types I and III universities over the 1977–87 period, there are no significant differences. The better performance of counties with degree-granting medical and engineering universities relative to their control group of counties may reflect the fact that they are likely to have a large number of businesses in the early stages of the "product cycle," which are rapidly growing in terms of employment. See Vernon, "The Product Cycle Hypothesis."

8. Raw data are arranged in SAS-generated contingency tables. We calculated chi-square, likelihood ratio chi-square, Mantel-Haenszel chi-square, and phi and contingency coefficients. Full results are available from the authors.

Chapter 9

1. Total R&D spending by the government, as a percentage of GNP, remained stable for the last half of the 1980s after having expanded in the first half of the decade. The percentages have been 2.3 percent (1970), 2.2 percent (1975), 2.3 percent (1980), 2.6 percent (1983, 1984), 2.7 percent (1985), and 2.6 percent (1987). Nondefense R&D grew as a percentage of GNP in the late 1970s and stabilized after 1981. In dollar terms, nondefense R&D spending grew at an average annual rate of 22.35 percent from 1975 to 1985 but only at a 3 percent rate after that. U.S. Bureau of the Census, *Statistical Abstract of the United States*, p. 578.

2. There are many references to this in the economic geography literature. See, for example, Massey, *Spatial Division of Labor*, and Dicken, "The Multi-Plant Business Enterprise and Geographic Space."

3. Cohen and Zysman, in *Manufacturing Matters*, argue that manufacturing still can provide the highest income per employee and rate of job growth relative to other uses of land.

4. See Chapter 1, n. 1.

5. See Luger, "The States and Industrial Development."

Bibliography

Alexander, T. "A Park That Reversed a Brain Drain." *Fortune* 95 (1977): 148–53.

Allen, David N., and David J. Hayward. "The Role of New Venture Formation/Entrepreneurship in Regional Economic Development." *Economic Development Quarterly* 4, no. 1 (February 1990): 55–63.

Allesch, J., and H. Fiedler, eds. *Management of Science Parks and Innovation Centers.* Berlin: 1985.

Andersson, Åke. "Creativity and Regional Development." *Papers of the Regional Science Association* 56 (1985): 5–20.

Appold, Steven. "The Geography of R&D Labs in the U.S." Unpublished manuscript, Department of Sociology, University of North Carolina, Chapel Hill, 1987.

Bania, Neal R., Robert Eberts, and Michael Fogarty. "The Role of Technical Capital in Regional Growth." Paper presented at the Western Economic Association Meetings, July 1987.

Berry, Brian J. L. *Growth Centers in the American Urban System.* Cambridge, Mass.: Ballinger, 1973.

———. "Hierarchical Diffusion: The Basis of Development Filtering and Spread in a System of Growth Centers." In *Growth Centers in Regional Economic Development*, edited by Niles Hansen, pp. 108–38. New York: The Free Press, 1972.

Binyon, M. "Research Parks: Where the Grass Is Always Greener." *The Times Higher Education Supplement*, September 9, 1977.

Bluestone, Barry, Bennett Harrison, and William Gorham. "Storm Clouds on the Horizon." In *The State and Local Industrial Policy Question*, edited by Harvey A. Goldstein, pp. 16–33. Chicago: APA Planners Press, 1987.

Bollinger, Lynn, Katherine Hope, and James Utterback. "A Review of Literature and Hypotheses on New Technology-Based Firms." *Research Policy* 12 (1983): 1–14.

Boudeville, J. R. *Problems of Regional Economic Planning.* Edinburgh: Edinburgh University Press, 1966.

Britton, J. N. H. "Environmental Adaptation of Industrial Plants: Service Linkages, Locational Environment and Organization." In *Spatial Perspectives on Industrial Organization and Decision-Making*, edited by F. E. I. Hamilton, pp. 363–90. New York: Wiley, 1974.

Brown, Wayne S. "Locally-Grown High Technology Business Development: The Utah Experience." In *Entrepreneurship and Technology: World Experiences and Policies*, edited by W. Brown and R. Rothwell, pp. 177–83. Harlow, Essex: Longmans, 1987.

Browning, J. E. *How to Select a Business Site.* New York: McGraw-Hill, 1980.

Business Week 20 (December 1952).

California Department of Health Services, Toxic Substances Control Division. *Stanford Research Park Sites*. California Department of Health Services, August 1988.

Cohen, Stephen S., and John Zysman. *Manufacturing Matters: The Myth of the Post-Industrial Economy*. New York: Basic Books, 1987.

Cook, Thomas D., and Donald T. Campbell. *Quasi-Experimentation*. Boston: Houghton Mifflin, 1979.

Cooper, A., and J. Komives. *Entrepreneurship: A Symposium*. Milwaukee, Wis.: Center for Venture Management, 1972.

Cox, D. R. "Regression Models and Life Tables." *Journal of the Royal Statistical Society* (1972): 187–220.

Cox, D. R., and R. Oakes. *Analysis of Survival Data*. London: Chapman and Hall, 1984.

Cox, R. N. "Lessons from 30 Years of Science Parks in the U.S.A." In *Science Parks and Innovation Centres: Their Economic and Social Impact*, edited by J. M. Gibb, pp. 17–25. Amsterdam: Elsevier Science Publications, 1985.

Danilov, Victor. "How Successful Are Science Parks?" *Industrial Research* 9 (May 1967): 76.

Darwent, D. F. "Growth Poles and Growth Centers in Regional Planning: A Review." *Environment and Planning* 1, no. 1 (1969): 5–31.

Dicken, P. "The Multi-Plant Business Enterprise and Geographic Space: Some Issues in the Study of External Control." *Regional Studies* 10 (1976): 401–12.

Erickson, Rodney A. "The Spatial Pattern of Income Generation in Lead Firm, Growth Area Linkage Systems." *Economic Geography* 51 (1975): 17–26.

Faculty-Staff Newsletter. Stanford, Calif.: Stanford University, March 28, 1960.

Fairbanks, R. "North Carolina Eyes State's 'Golden Goose'." *Los Angeles Times*, March 15, 1981.

Feller, Irwin. "Universities as Engines of R&D-Based Economic Growth: They Think They Can." *Research Policy* 19, no. 4 (1990): 335–48.

———. "University-Industry R&D Relationships." In *Growth Policy in the Age of High Technology*, edited by Jurgen Schmandt and Robert Wilson, pp. 313–43. Boston: Unwin Hyman, 1990.

Florax, Raymond, and H. Folmer. "Regional Economic Effects of Universities: The Impact of Knowledge Production on Investments of Industry." Paper presented at the 1989 Meetings of the Regional Science Association, Santa Barbara, Calif., November.

Franco, Michael R. "Key Success Factors for University-Affiliated Research Parks." Ph.D. dissertation, University of Rochester, 1985.

Friedmann, John. *Regional Development Policy: A Case Study of Venezuela*. Cambridge: MIT Press, 1985.

Geiger, Roger L. *To Advance Knowledge: The Growth of American Research Universities, 1900–1940*. New York: Oxford University Press, 1986.

Gibb, J. M., ed. *Science Parks and Innovation Centres: Their Economic and Social Impact*. Amsterdam: Elsevier Science Publications, 1985.

Gibson, Weldon B. *Stanford Research Institute: A Story of Scientific Service to Business, Industry, and Government*. New York: The Newcomen Society in North America, 1968.

Gilmour, J. M. "External Economies of Scale, Interindustry Linkages and Decision-Making in Manufacturing." In *Spatial Perspectives on Industrial Organization and Decision-Making*, edited by F. E. I. Hamilton, pp. 335–62. New York: Wiley, 1974.

Glasmeier, Amy. *The Making of High Tech Regions*. Princeton: Princeton University Press, 1990.

Glazer, Sarah. "Business Takes Root in University Parks." *High Technology* 6 (1986): 40–47.

Goldman, Robert, and Paul Luebke. "Corporate Capital Moves South: Competing Class Interests and Labor Relations in North Carolina's 'New' Political Economy." *Journal of Political and Military Sociology* 13 (Spring 1985): 17–32.

Goldstein, Harvey A., and Michael I. Luger. "Research Parks: Do They Stimulate Regional Economic Development?" *Economic Development Commentary* 13 (Spring 1989): 3–9.

———. "Universities and Regional Economic Development in the 1990s." Paper presented at the annual conference of the Association for Policy Analysis and Management, San Francisco, October 19, 1990.

Gore, Charles. *Regions in Question: Space, Development Theory and Development Policy*. London: Methuen, 1984.

Gross, J., and V. Clark. *Survival Distributions: Reliability Applications in the Biomedical Sciences*. New York: Wiley, 1975.

Hamilton, W. B. "The Research Triangle of North Carolina: A Study in Leadership for the Common Weal." *South Atlantic Quarterly* 65 (Spring 1966): 254–78.

Hansen, Niles. "Development of Pole Theory in a Regional Context." *Kyklos* 20 (1967): 709–25.

Harris, Candace S. "Establishing High-Technology Enterprises in Metropolitan Areas." In *Local Economies in Transition*, edited by Edward M. Bergman, pp. 165–84. Durham, N.C.: Duke University Press, 1986.

Higgins, Benjamin. "From Growth Poles to Systems of Interaction in Space." *Growth and Change* (October 1983): 7.

Hirschman, Albert O. *The Strategy of Economic Development*. New Haven: Yale University Press, 1958.

Holland, Stuart. *Capital versus the Regions*. New York: St. Martin's Press, 1976.

Howard, Ebenezer. *The Garden City Movement*. Hitchin, England: Garden City Press, 1905.

Hull, McAllister H., Jr. "The University and Economic Development." *Journal of Social, Political, and Economic Studies* 10 (Winter 1985): 481–87.

Jaffe, Adam B. "Academic Research with Real Effects." Working Paper, Center for Regional Issues, Case Western Reserve University, 1989.

———. "Technological Opportunity and Spillovers of R&D: Evidence from Firms'

Patents Profits and Market Value." *American Economic Review, Papers and Proceedings* 76, no. 2 (1986): 984–1001.

Junne, G. "Der Strukturpolitische Wettlauf zwischen den Kapitalistische Industrielandern." *Politische Vierteljahresschrift* 25 (1984): 138.

Kalbfleisch, J., and R. Prentice. *The Statistical Analysis of Failure Time Data.* New York: Wiley, 1980.

Klausner, Samuel Z. "Dilemmas of Usefulness: Universities and Contract Research." Paper presented to the American Bar Foundation, April 18, 1988.

Krenz, Charles. "Silicon Valley Spin-Offs from Stanford University Faculty." Unpublished report. Stanford, Calif.: Stanford University Business School, August 1988.

Krist, H. "Innovation Centres as an Element of Strategies for Endogenous Regional Development." In *Science Parks and Innovation Centres: Their Economic and Social Impact*, edited by J. M. Gibb, pp. 178–89. Amsterdam: Elsevier Science Publications, 1985.

Kuttner, Robert. "The Declining Middle." *Atlantic Monthly* 252 (July 1983): 60–69.

Lasuén, J. R. "Multi-Regional Economic Development: The Open System Approach." In *Information Systems for Regional Development: A Seminar*, edited by T. Hagerstrand and A. Kuklinski, pp. 169–212. Lund Studies in Geography, series B, Human Geography 37. Lund: Gleerup, 1971.

———. "Urbanization and Development—The Temporal Interaction between Geographical and Sectoral Clusters." *Urban Studies* 10 (1973): 163–88.

Levitt, Rachelle, ed. *The University/Real Estate Connection: Research Parks and Other Ventures.* Washington, D.C.: The Urban Land Institute, 1987.

Little, William F. "Research Triangle Park." *The World & I* (November 1988): 178–85.

Los Altos News, January 14, 1955.

Lowood, Henry. Letter to one of the authors, August 30, 1989.

———. "From Steeples of Excellence to Silicon Valley." *Campus Report*, March 9, 1988.

Luebke, Paul. "Corporate Conservatism and Government Moderation in North Carolina." In *Perspectives on the American South*, edited by Merle Black and John Shelton Reed, pp. 107–18. New York: Goldman and Breech, 1981.

Luebke, Paul, Stephen Peters, and John Wilson. "The Political Economy of Microelectronics in North Carolina." In *High Hopes for High Tech: The Microelectronics Industry in North Carolina*, edited by Dale Whittington, pp. 1310–28. Chapel Hill: University of North Carolina Press, 1985.

Luger, Michael I. "The States and Industrial Development: Program Mix and Policy Effectiveness." In *Perspectives on Local Public Finance and Public Policy*, edited by John M. Quigley, pp. 29–64. Greenwich, Conn.: JAI Press, 1987.

McDonough, Steven. "An Empirical Analysis of Suburban Industrial Park Development and Industrial Site Prices." Ph.D. dissertation, University of Pennsylvania, 1980.

Malecki, Edward J. "Dimensions of R&D Location in the United States." *Research Policy* 9 (1980): 2–22.

———. "Recent Trends in the Location of Industrial Research and Development: Regional Development Implications for the United States." Unpublished manuscript, University of Florida, n.d.

———. "The R&D Location Decision of the Firm and 'Creative Regions': A Survey." *Technovation* 6 (1987): 205–22.

Malizia, Emil. "The Locational Attractiveness of the Southeast to High-Technology Manufacturers." In *High Hopes for High Tech: The Microelectronics Industry in North Carolina*, edited by Dale Whittington, pp. 173–92. Chapel Hill: University of North Carolina Press, 1985.

Manski, C., and D. McFadden, eds. *Structural Analysis of Discrete Data with Econometric Applications.* Cambridge: MIT Press, 1981.

Markusen, Ann R. *Profit Cycles, Oligopoly, and Regional Development.* Cambridge: MIT Press, 1985.

Markusen, Ann, Peter Hall, and Amy Glasmeier. *High Tech America: The What, How, Where, and Why of the Sunrise Industries.* Boston: Allen and Unwin, 1986.

Massey, Doreen. *Spatial Division of Labor: Social Structures and the Geography of Production.* New York: Athuen, 1984.

Miller, Roger, and Marcel Cote. "Growing the Next Silicon Valley." *Harvard Business Review* (July/August 1985): 114–23.

Monck, C. S. P., et al. *Science Parks and the Growth of High Technology Firms.* London: Croom Helm, 1988.

Money, Mark L. "University Related Research Parks." Ph.D. dissertation, University of Utah, 1970.

Moseley, Malcolm J. *Growth Centres in Spatial Planning.* Oxford: Pergamon Press, 1975.

Murphy, Zera. Speech to the Second International Conference of the Association of University-Related Research Parks, Research Triangle Park, N.C., May 1987. (Available from Z. Murphy, Department of Lands Management, Encina Hall, Stanford University.)

Myrdal, Gunnar. *Economic Theory in Underdeveloped Regions.* London: Duckworth, 1957.

National Association of Realtors. *Commercial Real Estate Newsletter.* November 1989.

National Planning Association. *Regional Economic Projections Series, 1987.* 1988.

National Science Board. *University-Industry Research Relationships.* Washington, D.C.: U.S. Government Printing Office, 1983.

Nelson, Richard. "Institutions Supporting Technical Advances in Industry." *American Economic Review, Papers and Proceedings* 76, no. 2 (1986): 188.

Norton, R. D., and John Rees. "The Product Cycle and the Spatial Decentralization of American Manufacturing." *Regional Studies* 13, no. 2 (August 1979): 141–51.

Noyelle, Thierry. "Advanced Services in the System of Cities." In *Local Economies in Transition*, edited by Edward M. Bergman, pp. 143–65. Durham, N.C.: Duke University Press, 1986.

Oakey, R. P. *High Tech Industry and Industrial Location*. Hampshire, England: Gower, 1981.

O'Connor, Michael. "Mutual Benefits Boost University-Related Research Parks." In *Site Selection Handbook, 1981*, pp. 1281–84. Atlanta: Conway Publications, 1986.

Palo Alto Times, January 24, December 5, 1956.

Patenaude, Joel. "High-Tech Businesses in Durham County: Reasons for Their Location." Thesis submitted to the Department of City and Regional Planning, University of North Carolina, Chapel Hill, April 1987.

Perroux, François. "Economic Space, Theory, and Applications." *Quarterly Journal of Economics* 64 (1950): 89–104.

Pred, Allan. "The Interurban Transmission of Growth in Advanced Economies: Empirical Findings versus Regional Planning Assumptions." *Regional Studies* 10 (1976): 151–71.

Premus, Robert. *Location of High Technology Firms and Regional Economic Development*. Washington, D.C.: U.S. Government Printing Office, 1982.

Rees, John. "The Diffusion of New Production Technology: Implications for State and Local Industrial Policy." In *The State and Local Industrial Policy Question*, edited by Harvey A. Goldstein, pp. 60–72. Chicago: APA Press, 1987.

Rees, John, R. Briggs, and D. Hicks. "New Technology in the United States' Machinery Industry." In *The Regional Economic Impact of Technological Change*, edited by A. T. Thwaites and R. P. Oakley, 164–94. London: Frances Pinter, 1985.

Rees, John, and Howard Stafford. "A Review of Regional Growth and Industrial Location Theory: Towards Understanding the Development of High-Technology Complexes in the United States." Paper prepared for the Office of Technology Assessment, U.S. Congress, 1985.

Roberson, James O. Letter to the authors, August 24, 1989.

Sampson, Gregory. "Employment and Earnings in the Semiconductor Electronics Industry: Implications for North Carolina." In *High Hopes for High Tech: The Microelectronics Industry in North Carolina*, edited by Dale Whittington, pp. 256–95. Chapel Hill: University of North Carolina Press, 1985.

San Francisco Examiner, January 26, 1953.

Saxenian, Annalee. "The Genesis of Silicon Valley." In *Silicon Landscapes*, edited by Peter Hall and Ann Markusen. London: Allen and Unwin, 1985.

———. *Silicon Chips and Spatial Structure: The Industrial Basis of Urbanization in Santa Clara County, California*. Working Paper 345. Berkeley: Institute of Urban and Regional Development, University of California, 1981.

———. "The Urban Contradictions of Silicon Valley." In *Sunbelt-Snowbelt: Urban Development and Regional Restructuring*, edited by William Tabb and Larry Sawers, pp. 163–201. New York: Oxford University Press, 1984.

Schamp, E. W. "Technology Parks and Interregional Competition in the Federal

Republic of Germany." In *New Technology and Regional Development*, edited by G. A. van der Knapp and E. Weaver, pp. 120–31. London: Croom Helm, 1987.

Scott, Alan, and Michael Storper. "High Technology Industry and Regional Development: A Theoretical Critique and Reconstruction." *International Social Science Journal* 112 (May 1987): 215–33.

Sellars, Linda. "An Oral History of the Research Triangle Park." Ph.D. dissertation, University of North Carolina, Chapel Hill, 1991.

Shapero, Albert. "Entrepreneurship: Key to Self-Renewing Economies." *Commentary* 5 (April 1981): 19–23.

Sirbu, Marvin A., Jr., et al. *The Formation of a Technology Oriented Complex: Lessons from North American and European Experience*. Cambridge: MIT Center for Policy Alternatives, 1976.

Southern Growth Policies Board. *Halfway Home and a Long Way to Go: The Report of the 1986 Commission on the Future of the South*. Research Triangle Park, N.C.: SGPB, 1986.

Steinnes, Donald N. "On Understanding and Evaluating the University's Evolving Economic Development Policy." *Economic Development Quarterly* 1, no. 3 (August 1987): 214–44.

Sternberg, Rolf. *Technologie- und Gründerzentren als Instrument Kommunaler Wirtschaftsförderung*. Dortmund: Dortmunder Vertrieb für Bau- und Planungsliteratur, 1988.

Stöhr, Walter. "The Spatial Dimension of Technology Policy: A Framework for Evaluating the Systemic Effects of Technological Innovation." Working Paper, Wirtschaftsuniversität Wien, Vienna, 1987.

Thomas, M. D. "Regional Economic Development and the Role of Innovation and Technical Change." In *The Regional Economic Impact of Technological Change*, edited by A. T. Thwaites and R. P. Oakley, pp. 13–39. London: Frances Pinter, 1985.

Thwaites, A. T., A. Edwards, and D. Gibbs. *Interregional Diffusion of Product Innovations in Great Britain*. Final Report to the Department of Industry and the EEC. Newcastle University: Centre for Urban and Regional Development, 1982.

Thwaites, A. T., R. P. Oakey, and P. A. Nash. "Technological Change and Regional Development: Some Evidence on Regional Variations in Product and Process Innovation." *Environment and Planning A* 14 (1982): 1073–86.

U.S. Bureau of the Census. *1950 Census of Housing*. Washington, D.C.: U.S. Government Printing Office, 1953.

———. *1980 Census of Housing*. Washington, D.C.: U.S. Government Printing Office, 1983.

———. *1950 Census of Population*. Washington, D.C.: U.S. Government Printing Office, 1953.

———. *1960 Census of Population*. Washington, D.C.: U.S. Government Printing Office, 1963.

————. *1970 Census of Population*. Washington, D.C.: U.S. Government Printing Office, 1973.

————. *1980 Census of Population*. Washington, D.C.: U.S. Government Printing Office, 1983.

————. *County Business Patterns*. Washington, D.C.: U.S. Government Printing Office, 1959, 1963.

————. *State and Metropolitan Data Book, 1986*. Washington, D.C.: U.S. Government Printing Office, 1988.

————. *Statistical Abstract of the United States*. Washington, D.C.: U.S. Government Printing Office, 1989.

U.S. Congress. Office of Technology Assessment. *Technology, Innovation, and Regional Economic Development, Background Paper #2: Encouraging High-Technology Development*. Washington, D.C.: Office of Technology Assessment, U.S. Congress, February 1985.

U.S. Department of Commerce. Bureau of Economic Analysis. *Local Area Personal Income*. Washington, D.C.: U.S. Government Printing Office, 1960–87 (data tapes, IRSS).

————. *RIMS II*. Washington, D.C.: U.S. Government Printing Office, 1986.

Vernon, Raymond. "International Investment and International Trade in the Product Cycle." *Quarterly Journal of Economics* 80 (1966): 190–207.

————. "The Product Cycle Hypothesis in an International Environment." *Oxford Bulletin of Economics and Statistics* 41, no. 4 (1979): 255–67.

Vinson, R., and P. Harrington. *Defining "High Technology" Industries in Massachusetts*. Boston: Department of Manpower Development, Commonwealth of Massachusetts, 1979.

Vogel, Ezra. *Comeback: Building the Resurgence of American Business*. New York: Touchstone, 1985.

von Blum, Paul. *Stillborn Education: A Critique of the American Research University*. Lanham, Md.: University Press of America, 1986.

Weiss, Julian. "High-Tech Facilities: The Verdict Is Still Out but R&D Parks Catch On." *Business Facilities* 20 (February 1987): 69–76.

"What's Happened Down on the Farm?" *Recap* 15 (March–April 1958): 2.

Williams, J. C. "University-Industry Interactions: Finding the Balance." *Engineering Education* 77 (March 1986): 320–25.

Wilson, Louis R. *The Research Triangle of North Carolina*. Chapel Hill, N.C.: Louis R. Wilson, 1967.

Index

DATE DUE

Ill. 12-9-93			